Review of African Granulites and Related Rocks

TOM N. CLIFFORD
Department of Geology
University of the Witwatersrand
Johannesburg, South Africa

THE
GEOLOGICAL SOCIETY
OF AMERICA

SPECIAL PAPER 156

WILLIAM MADISON RANDALL LIBRARY UNC AT WILMINGTON

Copyright 1974 by The Geological Society of America, Inc.
Library of Congress Catalog Card Number 74-84196
I.S.B.N. 0-8137-2156-3

Published by
THE GEOLOGICAL SOCIETY OF AMERICA, INC.
3300 Penrose Place
Boulder, Colorado 80301

Printed in the United States of America

QE475
.G7
.C55

The printing of this volume has been made possible through the bequest of Richard Alexander Fullerton Penrose, Jr., and the generous support of all contributors to the publication program.

245922

Contents

Abstract

Although granulite facies rocks are present in zones of orogenesis ranging in age from 300 to 3,000 m.y. in Africa, only two distinct ages of granulite facies metamorphism, at about 1,200 and 3,000 m.y., have thus far been recognized. Granulite facies suites in regions of orogenesis of other ages — 1,850 ± 250 m.y., 600 ± 100 m.y., and Phanerozoic — represent older remnants since affected by regional retrograde metamorphism. The retrogression is particularly widespread in the 600 ± 100 m.y. orogenic zone, an important part of which is interpreted in terms of a former zone of granulite facies rocks, remnants of which are now preserved as isolated masses in amphibolite facies rocks.

In most African examples, hypersthene granulites of silicic to mafic and ultramafic composition are associated with undoubted metasedimentary rocks and a variety of other rock types, including anorthosite; intrusive charnockite also occurs in a number of regions. The alkali-silica plot for 240 analyses of pyroxene granulite and charnockite suggests a chemical affinity with high-alumina basalt and its calc-alkaline derivatives; the AFM trends are similar to those of calc-alkaline and alkaline igneous suites.

With few exceptions, rocks of appropriate composition in the granulite facies contain hypersthene + diopside + plagioclase ± garnet, reflecting medium to high pressure. Available Sr-isotopic data for granulite-charnockite suites of Africa and elsewhere in the world yield a crude "evolution path" that may reflect Sr-isotopic development in the deeper parts of the Earth's crust or in the mantle beneath the continental crust.

Introduction

The early crustal history of Africa has been discussed in summary works by Nicolaysen (1962), Kennedy (1965), Nicolaysen and Burger (1965), Cahen and Snelling (1966), Black (1967), Clifford (1968, 1970), and Black and Girod (1970). On the basis of available geologic and geochronologic data, it has been argued (Clifford, 1972) that the major part of the continent is a segment of primeval crust more than 3,000 m.y. old, which has been affected by several major orogenies; these include at least two major orogenic events older than 2,500 m.y., for which the best evidence is preserved in a number of pristine nuclei (Fig. 1, nos. 1 through 8), and a sequence of polyepisodic orogenies at 1,850 ± 250 m.y. ago (Eburnian and Huabian orogenic episodes), 1,100 ± 200 m.y. ago (Kibaran orogeny), 600 ± 100 m.y. ago (Damaran-Katangan or Pan-African orogeny) and during middle Paleozoic–early Mesozoic and Tertiary time (Fig. 1).

Recent data indicate the presence of ensialic floor beneath deformed geosynclinal cover in orogenic zones of all ages (Clifford, 1972), and in large segments of almost all of the orogenic zones, the orogenic event is entirely recorded as rejuvenated floor rocks. Metamorphic and intrusive rocks of the granulite-charnockite suite are concentrated in these floor segments (Clifford, 1973). Despite a voluminous literature on individual examples of these suites in Africa, few papers have dealt with their distribution and significance, although Oliver (1969) showed the geographic occurrence of some of the larger masses, and Saggerson and Owen (1969) commented on their significance as geobarometers. This review is intended to redress this situation, and with this aim in mind, semantic discussions of the terms "granulite" and "charnockite" are avoided here; the reader is referred to recent papers on this subject by Pichamuthu (1969) and Behr and others (1971). The term "granulite" is used in this paper for metamorphic rocks of appropriate texture in the granulite facies, and "charnockite" is used for the hypersthene-bearing intrusive suites as defined by Holland (1900); the adjective "charnockitic" is occasionally used for rocks whose field relations are still uncertain.

With the exception of the Alpine zone of North Africa, granulite-charnockite suites are present in major structural domains of all ages (Fig. 1). The ages of many of the suites differ from the dates of tectonism and metamorphism in those domains (see Table 1). The significance of the African suites is, therefore, considered here in terms of structural domains (Fig. 1): the ancient nuclei that have been unaffected by orogeny since ~2,500 m.y. ago (Fig. 1, nos. 1 through 8) are considered first; thereafter, suites in orogenic zones of successively younger age are discussed.

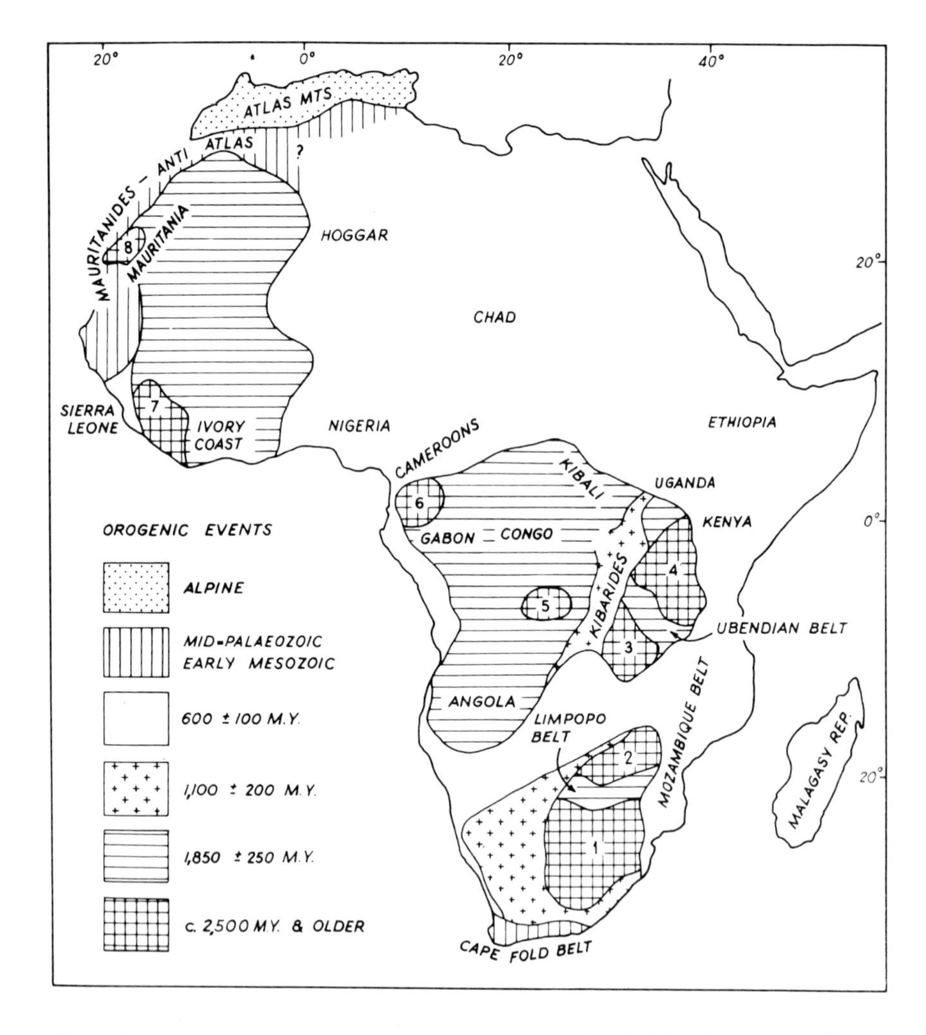

Figure 1. Generalized map of major orogenic structural units of Africa, showing zones in which individual orogenic events are preserved in their "primary" form, undisturbed by subsequent orogeny. Evidence of earliest events, believed to have affected the whole African crust, is now most clearly preserved in remnant "nuclei," numbered as follows: 1, Transvaal; 2, Rhodesia; 3, Zambia; 4, Dodoma-Nyanza; 5, Kasai; 6, Gabon-Cameroons; 7, Sierra Leone–Ivory Coast; 8, Mauritania.

TABLE 1. AGES OF AFRICAN CHARNOCKITE-GRANULITE SUITES IN RELATION TO STRUCTURAL UNITS IN WHICH THEY OCCUR*

Structural domain (and age of last major orogenesis)	Territory	Age of granulite-charnockite suite (m.y.)	Reference
	Congo	2,860 ± 120	Delhal and Ledent, 1971
		3,020 ± 125	
		2,865	
Ancient nuclei (>2,500 m.y.)	Ivory Coast	2,860 ± 140	Papon and others, 1968
		2,915 ± 115	
	Sierra Leone	3,020 ± 50	Andrews-Jones, 1968
	Cameroons	>2,490	Bessoles and Roques, 1959
	Gabon	>2,500	Bessoles and Roques, 1959
	Tanzania	2,280–2,750	Spooner and others, 1970
	Mauritania	3,090 ± 135	Vachette and Bronner, 1973, unpub. data
Eburnian-Huabian (1,850 ± 250 m.y.)	Tanzania	>2,100?	This paper
	Southern Africa (Limpopo Belt)	>2,690 ± 60	Van Breemen, 1970
	Gabon	>2,500	This paper
Kibaran (1,100 ± 200 m.y.)	South Africa (Namaqualand)	1,213 ± 22	Clifford and others, 1974
	South Africa (Natal)	>1,000	Nicolaysen and Burger, 1965
	Algeria (Hoggar)	3,030	Ferrara and Gravelle, 1966
	Uganda	2,655 ± 117	Spooner and others, 1970
		2,900	Leggo and others, 1971
	Rhodesia	>2,850 ± 50	Snelling, 1966
	Sierra Leone	⩾2,600	Hurley and others, 1971
		>1,340 ± 130	Allen and others, 1967
Damaran-Katangan or Pan-African (600 ± 100 m.y.)	Nigeria	>2,340 ± 70	Grant, 1970
		>635	Hurley and others, 1966
	Mozambique	>1,780	This paper
	Dahomey (Kouandé Gneiss)	>1,750 ± 230	Bonhomme, 1962
	Malawi	⩾1,126 ± 29	Clifford and others, in prep.
	Tanzania (Pare Mts.)	936 ± 63	Spooner and others, 1970
	Kenya	>855 ± 30	Cahen and Snelling, 1966
	Tanzania (Loibor Serrit)	731 ± 8	Spooner and others, 1970
	Central African Republic	>600	Roubault and others, 1965
Hercynian	Spanish Sahara	>3,000?	This paper

* See Figure 1.

Structural Distribution and Petrology

ANCIENT NUCLEI

All of the nuclei are characterized by belts of schist or greenstone, which are intruded by granitic plutonic rocks. Within these pristine remnants of the African crust, granulite-charnockite suites have been described from the Kasai, Sierra Leone–Ivory Coast, and Gabon-Cameroons nuclei (Fig. 2); other examples are recorded from the margin of the Dodoma-Nyanza nucleus, and possible relatives of these rock types have been described from the Rhodesia and Transvaal nuclei and from the northern Congo.

Kasai

The Kasai nucleus (Fig. 2, no. 5) consists of a suite of widespread metamorphic and plutonic rocks unconformably overlain by younger sedimentary and volcanic rocks of the Luiza and Lulua Series (Fig. 3). Of particular concern in this work is the extensive Lulua Massif of noritic gabbro and granulite facies rocks; Beugnies (1953) described a similar suite in Katanga that may represent the eastward continuation of the Lulua suite.

In the Lulua Massif, Delhal (1957, 1958, 1963) recognized three zones: a noritic gabbro complex, composed of noritic gabbro, garnet gabbro, and amphibolite, in the north (Fig. 3); a charnockitic complex, composed of hypersthene-quartz-feldspar-biotite gneiss, plagioclase-quartz-hypersthene (enderbitic) gneiss, quartz–K-feldspar–hypersthene (charnockitic) gneiss, garnet-sillimanite gneiss, and leptynite, in the south; and a transitional zone of rocks of quartz diorite composition between these two contrasted suites (Delhal, 1957, p. 233). The hypersthene-bearing gneisses clearly reflect granulite facies metamorphic conditions in the charnockitic complex. In contrast, Delhal (1963, p. 14) considered that the noritic gabbros are of magmatic origin, emplaced under the physical conditions of granulite facies metamorphism. After these high-temperature events, retrograde metamorphism resulted in the development of garnet and (or) hydrous minerals, notably amphibole, mica, chlorite, and zoisite in (1) the so-called amphibolo-gneissic zone and (2) the Lueta Complex (Fig. 3).

Recent age determinations for zircon and monazite from charnockitic and enderbitic gneisses of the Lulua Massif (Delhal and Ledent, 1971) have given a U-Pb age of 2,865 m.y. (Table 1). The Rb-Sr whole-rock isochron age of these rocks, together with a number of leucocratic gneiss and garnet gneiss samples, is 2,860 ± 120 m.y.[1] (Delhal and Ledent, 1971, p. 217). However, the isochron age of leucocratic gneiss samples is 3,020 ± 125 m.y., and this is considered to be the age of granulite facies metamorphism (Delhal and

[1] All Rb-Sr ages refer to a Rb^{87} decay constant of $1.39 \times 10^{-11}\ yr^{-1}$.

Ledent, 1971). Mineral ages in the range 2,000 to 2,700 m.y. reflect the subsequent retrogressive metamorphism (Delhal and Ledent, 1965, p. 109; 1971).

Sierra Leone–Ivory Coast

The Sierra Leone–Ivory Coast nucleus (Fig. 2, no. 7) consists largely of granitic rocks, together with metasedimentary and metavolcanic rocks that have been ascribed to the Birrimian System in the Ivory Coast (Bolgarsky, 1950) and to the Kambui Schist in Sierra Leone (Haughton, 1963). Samples of Kambui Schist, granitic gneiss, and granite have given a Rb-Sr isochron age of ~2,800 m.y. (Hurley and others, 1971).

Lacroix (1910) first described rocks of charnockitic affinities from this part of Africa, from the Man Massif in the Ivory Coast. Similar rocks have since been described from Sierra Leone (Wilson, 1965; Andrews-Jones, 1968). The Man Massif is mainly composed of

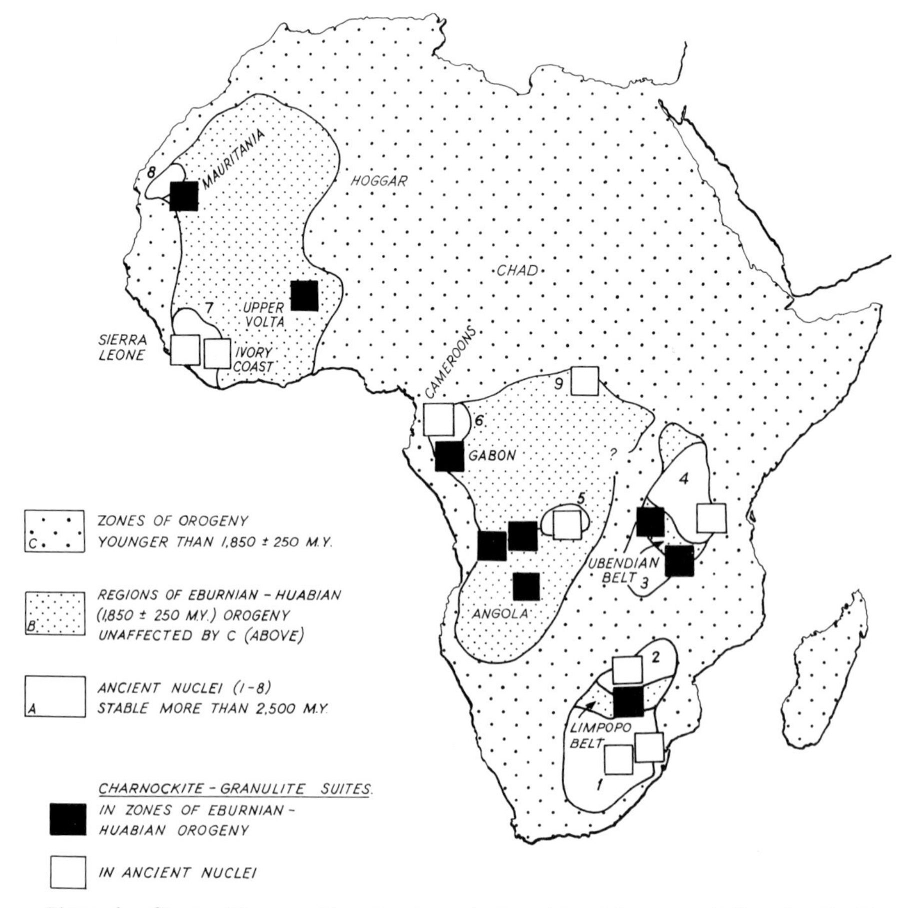

Figure 2. Charnockite-granulite suites in ancient nuclei and in zones of Eburnian-Huabian orogenesis. Numbers refer to descriptions in text.

a suite of plutonic rocks (Marvier, 1953), most of them hypersthene granite and norite; additionally, leucocratic norite and melanocratic rocks, including hypersthenite and pyroxenite, are present. Both Legoux (1939) and Bolgarsky (1950) argued that this suite postdated the Birrimian System, and the metamorphic influence of the plutonism is shown by the presence of hypersthene and almandine-pyrope garnet in the metasedimentary and metavolcanic rocks (Legoux, 1939, p. 70–76). Recent geochronologic studies of whole-rock samples of the hypersthene rocks and associated rock types have yielded two Rb-Sr isochrons that define comparable ages of 2,860 ± 140 m.y. and 2,915 ± 115 m.y., but the studies yielded two distinct Sr^{87}/Sr^{86} initial ratios of 0.699 ± 0.001 and 0.707 ± 0.001, respectively (Papon and others, 1968). These ratios are interpreted in terms of two different strontium sources: that for the 2,860 ± 140 m.y. samples at depth, and that for the 2,915 ± 115 m.y. samples in the sialic crust (Papon and others, 1968).

In Sierra Leone, Wilson (1965) described a suite of granulite facies rock, the Mano Moa granulite, occurring as lenslike masses within the extensive group of granitic rocks that characterizes most of eastern Sierra Leone. This granulite suite includes hypersthene-garnet gneiss, mafic and ultramafic granulite containing hypersthene and clinopyroxene, hypersthene-clinopyroxene-garnet–bearing quartz-feldspathic granulite, anorthosite, quartzite, and ferruginous gneiss containing hypersthene and magnetite. Andrews-Jones (1968) demonstrated the progressive transformation of amphibolite to pyroxene amphibolite to pyroxene gneiss with increasing metamorphic grade in the Kambui Schist, and obtained a whole-rock Rb-Sr isochron age of 3,020 ± 50 m.y. for the areally associated granitic rocks, thus establishing a minimum age for the granitic suite and the metamorphism.

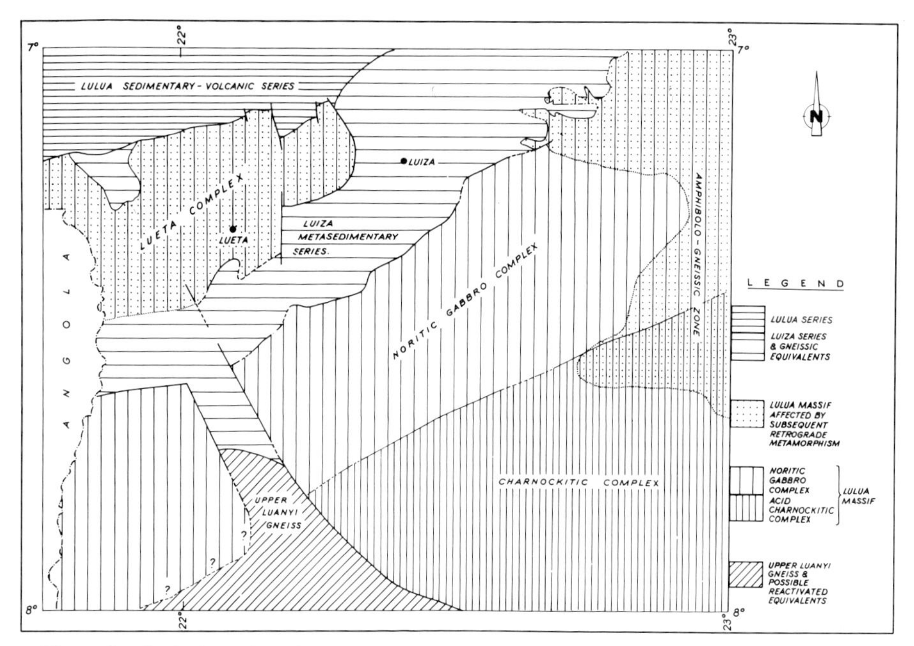

Figure 3. Geologic setting of noritic gabbro and charnockitic complexes of Lulua Massif, Kasai, Zaire (Congo). (After Delhal, 1957, 1963.)

Gabon-Cameroons

The largest area of charnockite-granulite rocks in the Gabon-Cameroons nucleus (Fig. 2, no. 6) is the Ntem Complex of the southern Cameroons. This complex has two components: a group of high-grade metamorphic rocks and an extensive suite of granitic plutonic rocks (Gazel and others, 1956). This crystalline suite extends southwestward into Spanish Guinea (Füster, 1958), and similar rock types are well developed in northern Gabon (Hudeley and Belmonte, 1970).

The metamorphic group of the Ntem Complex consists largely of augite-biotite-hypersthene gneiss, pyroxene amphibolite, quartzite, biotite gneiss, and a variety of retrograde rocks (Champetier de Ribes and Aubague, 1956; Champetier de Ribes and Reyre, 1959). The associated plutonic suite occupies approximately 50 percent of the complex and is predominantly composed of hypersthene and (or) diopside-bearing granite and granodiorite.

Three biotites from the plutonic suite of the Ntem Complex have yielded Rb-Sr ages of 2,000 to 2,400 m.y., whereas zircon from hypersthene granite gave Pb-α ages of 2,490 and 3,200 m.y. (Bessoles and Roques, 1959; Lasserre, 1964a, 1964b). It appears, therefore, that the granulite facies metamorphic rocks and their associated plutonic types are older than 2,500 m.y. and may include elements older than 3,000 m.y.

To the south, in Gabon, a number of small occurrences of charnockite-granulite form part of a suite of gneiss, leptynite, and quartzite (Aubague and Hausknecht, 1957, 1959). The pyroxene granulite includes clinopyroxene-bearing and, less commonly, hypersthene-bearing types, which form a large massif in the Mitzic district. A minimum age for the granulite facies rocks is provided by a Pb-α age of 2,500 m.y. for zircon from intrusive granite in northern Gabon (Bessoles and Roques, 1959).

Dodoma-Nyanza

The Dodoma-Nyanza nucleus (Fig. 2, no. 4) consists of two contrasted parts: (1) the major part of central Tanzania, underlain largely by granitic plutonic rock and associated metamorphic rock of the Dodoman System that are older than 2,300 to 2,500 m.y. (Cahen and Snelling, 1966); and (2) a northern part characterized by sedimentary and volcanic rocks of the Nyanzian and Kavirondian Systems and by granitic plutons.

Within the domain of the Dodoman System, Temperley (1938) identified charnockitic gneiss in eastern Tanzania. Geochronologic studies of these rocks have yielded a whole-rock Rb-Sr isochron age of 2,280 to 2,750 m.y. (Spooner and others, 1970). Biotite ages from this locality are K-Ar, 3,600 ± 300 m.y., and Rb-Sr, 475 ± 80 m.y. (Spooner and others, 1970).

Rhodesia

The Rhodesia nucleus (Fig. 2, no. 2) is characterized by remnant schist belts and large, engulfing bodies of granitic rock (Macgregor, 1951). The whole suite is older than 2,600 to 2,700 m.y., and the lower part of the schist-belt sucession is older than 2,900 m.y. (Wilson and Harrison, 1973). Stowe (1968, p. 66) identified a suite of ancient migmatites situated unconformably below the schist-belt succession. These migmatites contain relics of quartz-almandine-pyroxene-magnetite granulite and olivine-pyroxene granulite that *may* reflect granulite facies conditions.

Transvaal

The ancient rocks of the Transvaal nucleus (Fig. 2, no. 1) consist largely of intrusive suites of granitic rocks and schist belts of metasedimentary and volcanic rocks of the Swaziland System (Visser, 1956; Viljoen and Viljoen, 1970), which is certainly older than 3,000 m.y. and includes rocks as old as 3,400 ± 100 m.y. (Van Niekerk and Burger, 1969; Allsopp and others, 1969). In Swaziland, Hunter (1970) described a suite of gneiss that is considered to *predate* the Swaziland System. This Ancient Gneiss Complex includes cordierite-, garnet-, biotite-, and hornblende-bearing gneiss, together with quartzite, ironstone, amphibolite, and diopside granulite. Of particular interest is the presence of hypersthene-bearing quartz-magnetite-grunerite ironstones. Furthermore, J. V. Hepworth (1971, personal commun.) has noted the presence of charnockitic variants of a quartz diorite–granodiorite suite that is older than 2,900 m.y. and intrudes the Ancient Gneiss Complex. In South Africa, Willemse (1938, p. 46–54) recognized a number of granulite rock types in the Vredefort Dome. In particular, he noted garnet-feldspar-quartz rocks and hypersthene-garnet-magnetite-quartz rocks.

Lower Uelé, North Zaire (Congo)

Close to the northern border of Zaire (Fig. 2, no. 9), the Precambrian sequence includes a suite of gneiss of the Bomu Complex, believed to be older than 3,500 m.y. (Cahen and Lepersonne, 1967, p. 170). To date, no evidence has been presented to support a granulite facies metamorphism in this complex, but hypersthene granite is recorded from the region (Cahen and Lepersonne, 1967); for this reason, the Lower Uelé suite is noted here.

EBURNIAN-HUABIAN DOMAIN

Orogenic activity was widespread in southern and West Africa during the Eburnian and Huabian events, 1,850 ± 250 m.y. ago (see Fig. 1). With few exceptions (Clifford, 1972), this orogenic event is recorded, at the present erosion level, as tectonic and thermal reactivation of the lateral extensions of the schist belt and plutonic suites of the nuclei. Within these regions, representatives of granulite-charnockite suites occur notably in the Amsaga region of Mauritania; Angola; the Ubendian Belt of Tanzania and Malawi; the Limpopo Belt in Rhodesia, Transvaal, and Botswana; and a number of small bodies in Gabon and Upper Volta (Fig. 2). Nevertheless, no evidence of prograde granulite facies metamorphism of Eburnian-Huabian age has been noted; in all cases that have been studied by detailed geochronology, the granulite facies suites are remnants of older metamorphic events, subsequently subjected to retrograde metamorphism during Eburnian-Huabian time.

Mauritania

High-grade metamorphic rocks of the Amsaga Group were the subject of an important study by Blanchot (1954, 1955). More recently, Barrère (1967, 1969) showed the regional distribution of rocks of granulite and almandine amphibolite facies within this group, and he termed them the Saouda Series and Rag el Abiod Series, respectively. The Saouda Series includes a wide range of rock types: hypersthene–hornblende–K-feldspar–plagioclase–quartz gneiss, garnet leptynite, sillimanite-cordierite-garnet rock, pyroxene amphibolite, anorthosite, ferruginous quartzite, calc-silicate rock, and marble. In contrast, the Rag el Abiod Series is an extensive suite of almandine amphibolite facies rocks that is considered to represent the Saouda Series transformed, in large part, into porphyritic and migmatitic granite and migmatite (Barrère, 1967, 1969). The Rag el Abiod Series passes westward into metasedimentary, metavolcanic, and granitic rocks of the Mauritania nucleus (see Fig. 2). In this regard, although the Amsaga Group has been widely affected by the Eburnian event (Bonhomme, 1962), a recent whole-rock Rb-Sr isochron age of 3,090 ± 135 m.y. is regarded as the age of the end of the granulite facies metamorphism (M. Vachette and G. Bronner, 1973, unpub. data).

Ubendian Belt

The Ubendian Belt (Fig. 2) of southern Tanzania, northern Malawi, northern Zambia, and eastern Zaire is an extensive northwest-trending zone consisting largely of high-grade metamorphic rocks and intrusive granite (Cahen and Snelling, 1966). Although the age of the metamorphic suite is not known, it is older than 2,100 m.y. in the extension of the Ubendian Belt in eastern Zaire (Congo) (Cahen, 1970, p. 98).

Within the metamorphic suite, granulite facies rocks have been described from southern Tanzania and northern Malawi (McConnell, 1950; Sutton and others, 1954; Fitches, 1968). The detailed work by Sutton and others (1954) in Tanzania revealed three distinct zones: a central Katumbi zone of hypersthene-bearing rocks, separating a garnetiferous zone on the northeast from a granitic zone on the southwest. The first of these includes a variety of layered granulites mainly of intermediate chemical composition; silicic and mafic types are rare. Associated rock types are kyanite-, garnet-, and hypersthene-bearing quartzite, and biotite gneiss. Sutton and others (1954) favor the view that the granulites may be the prograde representatives of the rocks of the garnetiferous zone, characterized by metasedimentary rocks having the assemblage almandine-kyanite-quartz-hypersthene, and with hypersthene, hornblende, and garnet in rocks of mafic and ultramafic composition. Subsequent retrograde metamorphism resulted in the widespread development of silicic rocks in the granitic zone.

To the southeast, in northern Malawi, extensive areas of sillimanite and garnet gneiss of the upper amphibolite facies grade into the Jembia River Granulite (Fitches, 1968). The latter is mostly leucocratic granulite, including sillimanite-, cordierite-, and biotite-bearing varieties, mafic granulite composed of hornblende, plagioclase, diopsidic pyroxene and (or) hypersthene, and ferruginous quartzite (Fitches, 1968, 1970).

Limpopo Belt

The Limpopo Belt has many features in common with the Ubendian Belt. It extends from Botswana on the west almost to the Mozambique border on the east (Macgregor, 1952; Nicolaysen, 1962; Mason, 1969). The belt is divisible into three distinct zones: a central zone in which the structures run oblique to the overall trend of the belt and two marginal zones in which the strike is generally parallel to the trend of the belt (Cox and others, 1965). In the central zone in Botswana, Wakefield (1971) has noted the presence of an early granulite facies metamorphism associated with large-scale isoclinal folding, followed by amphibolite facies metamorphism. In Rhodesia, Robertson (1968) described an extensive suite of granulite facies rocks that apparently represent the culmination of prograde metamorphism that increases southward from the Rhodesia nucleus (Robertson, 1968, p. 127). In South Africa, detailed studies of the central zone (Söhnge and others, 1948) have demonstrated the presence of an ancient granitic basement and an overlying suite of metasedimentary rocks of the Messina Formation (Bahnemann, 1971). The Messina Formation consists of garnetiferous granite-gneiss, marble, ironstone, amphibolite, cordierite-sillimanite-biotite gneiss, plagioclase-pyroxene granulite, metaquartzite, anorthositic gneiss, and hornblendite (Bahnemann, 1971, 1972). Locally, the structurally underlying granitic gneiss has been partially remelted to produce the Singelele Granite. Summarizing the metamorphic history of the region, Bahnemann (1971, p. 16–22) noted that the mineralogy of the Singelele Granite and the biotite-garnet and cordierite gneiss fit with the amphibolite facies, whereas the plagioclase-pyroxene granulite suggests a granulite facies grade. Regarding the latter point, Chinner and Sweatman (1968) described enstatite-cordierite-sillimanite-quartz rocks reflecting (1) early enstatite-kyanite-quartz crystallization at temperatures in excess of 800°C and pressure greater than 10 kb, and (2) a later phase of cordierite-sillimanite paragenesis formed at similar temperatures and at pressures of 8 to 10 kb.

Van Breeman (1970) obtained a Rb-Sr isochron age of 2,600 ± 120 m.y. for dikes that cut the granulite facies rocks of the northern zone of the Limpopo Belt. In addition, a whole-rock Rb-Sr isochron age of 2,690 ± 60 m.y. has been obtained for the Singelele and related granites, whose anatectic mobilization occurred under amphibolite facies conditions; this is a minimum age for the granulite facies metamorphism (Table 2). In contrast, Rb-Sr ages for biotite and K-feldspar–plagioclase mineral pairs are concordant at 2,000 ± 70 m.y. over the whole belt (Van Breemen, 1970, p. 59) and are considered to reflect tectonism, including regional cataclasis.

TABLE 2. SEQUENCE OF EVENTS IN THE EARLY HISTORY OF THE LIMPOPO BELT

	Event	Age (m.y.)
VI	Cataclasis	~2,000
V	Emplacement of satellites of the Great Dyke	2,600 ± 120
VI	Amphibolite facies metamorphism and anatexis	2,690 ± 60
III	Granulite facies metamorphism	>2,700
II	Deposition of the Messina Formation ≡ schist-belt sequences of the Rhodesia and Transvaal nuclei	
I	Older basement (for example, Bulai granite gneiss)	

TABLE 3. ISOTOPIC AGES AND INITIAL Sr^{87}/Sr^{86} RATIOS (R_o) OF PRECAMBRIAN ROCKS IN SOUTHERN MALAWI

Nature of Event	Age (m.y.)	R_o (Sr^{87}/Sr^{86})
Pegmatite formation and emplacement of posttectonic intrusive rocks	493 ± 16	0.7069 ± 0.0009
Cataclasis and metamorphism	550?	
Folding, amphibolite facies metamorphism, and emplacement of syntectonic intrusive rocks	718 ± 25	0.7063 ± 0.0004
Deposition of upper Precambrian, subsequently removed by erosion		
Folding and granulite facies metamorphism	⩾1,126 ± 29	0.7055 ± 0.0014

Angola

Westward from the noritic gabbro and charnockitic complex of the Kasai nucleus, in Angola, are similar rocks (Fig. 2, no. 5; Delhal and Fieremans, 1964). The most western examples, around Salazar, were the subject of a petrochemical study by Ribeiro de Albuquerque and Figueiredo Gomes (1962), who recognized three petrographic groups: diorite and syenodiorite containing hypersthene, augite, biotite, and antiperthitic andesine; granulite composed of labradorite, diopsidic augite, and hypersthene; and hybrid rock composed of andesine, microcline, biotite, augite, and hypersthene.

The general alignment of the Kasai and Angola examples of these rocks suggests that they represent a charnockitic complex extending from Kasai for a distance of 1,200 km in an easterly direction (Delhal and Fieremans, 1964, p. 2667–2668). Delhal and Fieremans (1964) considered that these rocks are a continuous zone of ancient basement that has escaped subsequent (~2,000 m.y.?) granitization.

To the south, in central Angola, Tyrrell (1916) noted a number of charnockitic rocks, including silicic and intermediate compositional types. The age of these rocks is still unknown.

Gabon

In central and southern Gabon, many small occurrences of granulite facies rocks have been recognized. In particular, Choubert (1954) noted the presence of hypersthene and hypersthene-garnet rock and grunerite-garnet assemblages, together with gabbro, hypersthenite, pyroxenite, anorthosite, and cordierite gneiss (Choubert, 1954, p. 47–51). The age of these rocks is not known, but they are tentatively correlated with the suites older than 2,500 m.y. of the Ntem and Mitzic Massifs of the Gabon-Cameroons nucleus to the north.

KIBARAN DOMAIN

In East, Central, and southern Africa, there are a number of zones that were sites of orogenesis during the Kibaran orogeny, 1,100 ± 200 m.y. ago, including the Kibaride Belt of Central and East Africa, and the Namaqualand-Natal Belt of southern Africa (Figs. 1, 4). The Kibaride Belt is occupied by 10,000 m of geosynclinal sedimentary and volcanic rocks of the Kibara and correlative groups that rest on the eroded roots of the older Ubendian-Rusizi Belt and thus have a maximum age of ~2,100 m.y. (see Cahen, 1970). In the Namaqualand-Natal Belt to the south (Fig. 5), Nicolaysen and Burger (1965) have demonstrated the effects of events ~1,000 m.y. ago, but there are no extensive geosynclinal sequences of Kibara Group age. This region is particularly interesting because granulite facies metamorphism represents the culmination of *prograde* regional metamorphism and because charnockitic intrusive rocks of silicic and intermediate composition are present.

Namaqualand

The major part of Namaqualand is underlain by granite gneiss, together with remnants of the metavolcanic and metasedimentary rocks of the Kheis System that are older than 2,600 m.y. (Nicolaysen and Burger, 1965). The near-equality of ages of 900 to 1,100 m.y. yielded

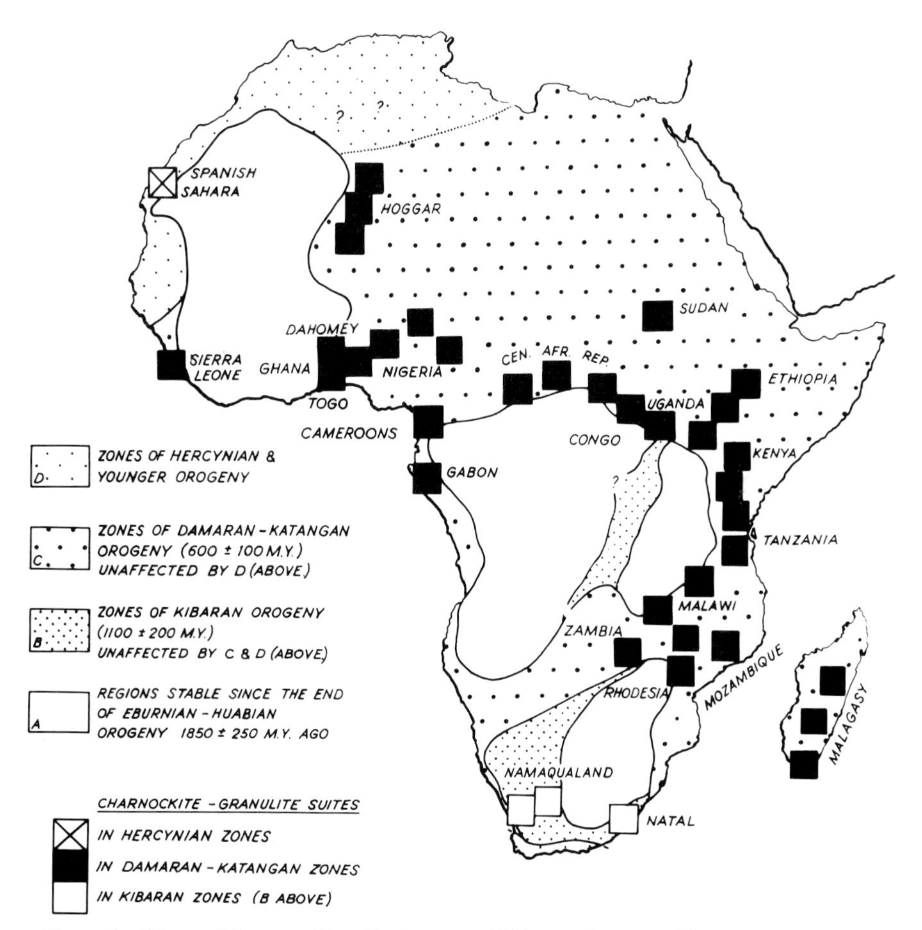

Figure 4. Charnockite-granulite suites in zones of Kibaran, Damaran-Katangan (Pan-African) and Hercynian orogenesis.

by a wide range of minerals is believed to reflect a period of intense reconstruction of older basement (Nicolaysen and Burger, 1965).

Within this vast region, several detailed and regional studies have been carried out on the regional metamorphism and structure (Rogers and Du Toit, 1908; Gevers and others, 1937; Coetzee, 1941, 1942; Poldervaart and Von Backström, 1949; De Villiers and Söhnge, 1959; Jansen, 1960; Von Backström, 1964, 1967; Benedict and others, 1964; Kroner, 1971; Joubert, 1971). These studies show that the regional metamorphic grade broadly increases southward from the Orange River, eastward from the Atlantic Coast, and southwestward from the northeastern part of Namaqualand (Fig. 5). Joubert (1971, p. 123) suggested that Namaqualand is the site of a major regional thermal dome culminating in granulite facies assemblages characterized by the coexistence of orthopyroxene and clinopyroxene, the occurrence of sillimanite-cordierite-garnet gneiss, and the local association clinopyroxene-almandine-plagioclase.

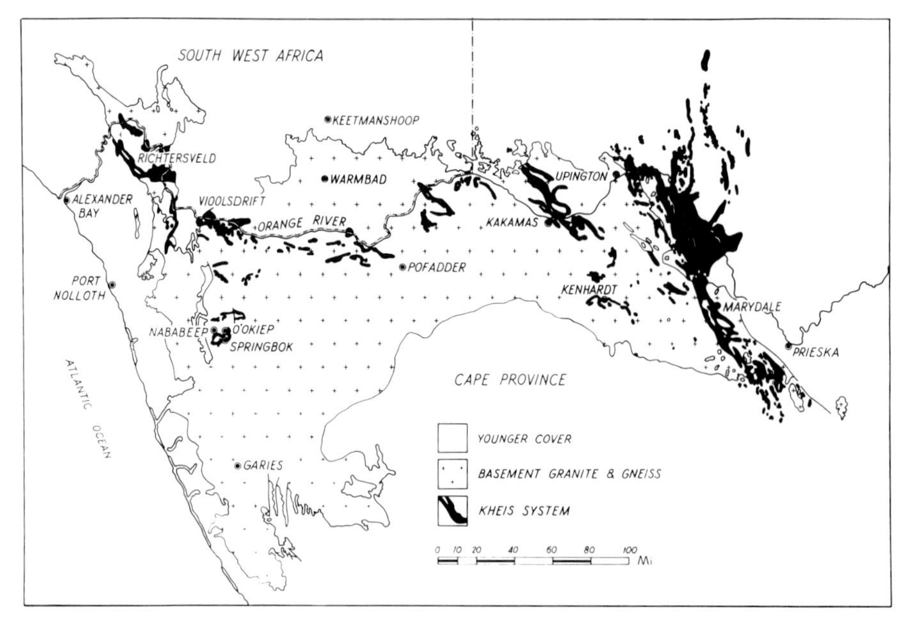

Figure 5. Basement geology of Namaqualand and southern South West Africa. (After Truswell, 1970.)

Of particular interest in a discussion of the high-grade metamorphic terrains is the Nababeep district (Fig. 5). Detailed studies (Benedict and others, 1964) have shown that the area is characterized by a suite of schist, quartzite, gneiss, and granulite, representing a structural thickness of more than 3,000 m (Benedict and others, 1964, p. 246). In this suite, several features are worthy of note: (1) primary muscovite is absent, and the essential mineralogy of the gneiss and granulite is quartz, microcline-microperthite, biotite, and plagioclase; (2) discontinuous layers of melanocratic rock contain hypersthene, diopside, hornblende, and plagioclase; (3) certain schists are characterized by sillimanite-garnet, cordierite-sillimanite, and cordierite-sapphirine-orthopyroxene assemblages; and (4) mafic intrusive rocks emplaced prior to the main metamorphism are now cordierite-hypersthene-anthophyllite rocks. The petrochemistry and age of a comprehensive suite of these metamorphic rocks have been studied (Clifford and others, 1974). The whole-rock Rb-Sr isochron age of 1,213 ± 22 m.y. (R_o = 0.7191 ± 0.0021) obtained from gneiss and granulite is considered to *reflect* the age of the progressive regional metamorphism.

Of additional interest is the presence of crosscutting bodies of hypersthene-bearing intrusive rocks, grouped locally as the noritoid suite, that include quartz diorite, diorite, norite, anorthosite, and hypersthenite. To the northeast a more granitic group of hypersthene-bearing charnockitic adamellite intrusive rocks has been described by Poldervaart and Von Backström (1949), Von Backström (1964), and Poldervaart (1966) from the Kakamas district (Fig. 5). In general, these rocks are unbanded and consist of quartz, plagioclase, K-feldspar, orthopyroxene and clinopyroxene, hornblende, and biotite. The contact effects of intrusion are reflected by the development of kinzigites that contain cordierite with or without sillimanite, corundum, andalusite, spinel, and staurolite (Von Backström, 1964); certain varieties contain hypersthene and garnet.

Field relations show that the emplacement of the charnockitic adamellite postdated the regional metamorphism. Nicolaysen and Burger (1965) obtained a Rb-Sr age of 1,105 m.y. for biotite from country-rock gneiss and comparable ages for zircon.

Natal

In the eastward extension of the Namaqualand Belt in Natal (Fig. 4), the Precambrian suite consists largely of (1) gneiss with some recognizable metasedimentary rocks, including calc-silicate marble, metasandstone that is now orthopyroxene-almandine granulite, and schist; and (2) intrusive rocks, including charnockitic types (Simpson and Tregidga, 1956; Gevers and Dunne, 1942; McIver, 1966). Among the intrusive rocks, a wide range of hypersthene-bearing rocks was recognized in the Port Edward region by Gevers and Dunne (1942) and McIver (1966), and they are intrusive into the metasedimentary Leisure Bay granulites, a group of hypersthene-garnet granulites and schists reflecting granulite facies metamorphism. The intrusive rocks include biotite diorite that contains plagioclase, alkali feldspar, clino-pyroxene, orthopyroxene, and quartz; fayalite-pyroxene adamellite composed largely of microcline perthite, plagioclase, quartz, orthopyroxene, hornblende, ferroaugite, olivine, and almandine; and orthopyroxene-bearing quartz diorite and granodiorite (McIver, 1966). Biotite from one of the youngest intrusive rocks has given a Rb-Sr age of 1,011 m.y. (Nicolaysen and Burger, 1965) and represents a minimum for the age of this intrusive suite.

DAMARAN-KATANGAN (PAN-AFRICAN) DOMAIN

A major part of the African continent was affected by Damaran-Katangan or Pan-African orogenesis, reflected by widespread radiometric ages of $\sim 600 \pm 100$ m.y. (Fig. 1), and the zones affected by this event can be subdivided into two types: (1) zones of orogenically deformed upper Precambrian geosynclinal sedimentary and volcanic rocks and (2) zones of rejuvenated older basement (Clifford, 1970, p. 13–14).

As in the Kibaran Belt, there is a clear spatial correlation between the zones of rejuvenated basement and the distribution of granulite-charnockite suites, notably in the Mozambique Belt in Mozambique, Malawi, Tanzania, Kenya, Ethiopia, and Sudan, and the westward extension of that zone in Uganda, northeastern Zaire (Congo), and the Central African Republic; in Togo, Dahomey, Ghana, and Nigeria; and in the Hoggar (see Figs. 1, 4). In addition, rocks of the granulite facies characterize the coastal zones of Sierra Leone, the Cameroons, and Gabon, and other examples are recorded from Zambia (Drysdall and others, 1972) and Rhodesia (Macgregor, 1951, p. lxii). Despite their structural distribution within the zone of 600 ± 100 m.y. orogenesis, the available geochronologic data indicate that most of these suites are remnants of metamorphic and plutonic events that significantly predated Damaran-Katangan tectonism and metamorphism (see Table 1). Moreover, in all regions that have been studied in detail, the granulite facies suites always show evidence of subsequent retrograde metamorphism.

Mozambique

Rocks of charnockite-granulite affinities are widely scattered in Mozambique (see Oberholzer, 1968). However, few of these have been studied in any detail, and perhaps the best known are in the Barué and Tete regions, where there are many minor occurrences in a country rock essentially composed of granitic gneiss (Coelho, 1954; Borges and Coelho, 1957; Assunçao and Coelho, 1956). The hypersthene-bearing rocks in the Barué area have been described by Araujo (1966, 1967), who recognized that they generally occur as small bodies or lenses of mafic granulite, diorite, tonalite, and pyroxenite associated with marble and quartzite. Araujo argued that this suite records a granulite facies metamorphism that predated the deposition of the Umkondo System, whose lower part is older than ~1,780 m.y. (Snelling, 1966). The Tete anorthosite-gabbro complex covers an area of 5,500 km^2 and consists of gabbro, anorthosite, pyroxenite, epidiorite, syenite, and granite (Freitas, 1957; Coelho, 1957, 1969; Vail, 1968). Davidson and Bennett (1950) noted that the gabbro is noritic, akin to charnockite. The complex has a minimum age of 580 m.y. (Darnley and others, 1961).

Malawi

Summary works by Bloomfield (1968) and Cannon and others (1969) show that Malawi is broadly divisible into a southern subprovince characterized by important occurrences of granulite facies rocks with associated meta-anorthosite and the development of marble and calc-silicate rock, and a northern subprovince in which these distinctive rock types are less well developed or are absent (Cannon and others, 1969).

The greater part of the metamorphic suite of the southern subprovince is characterized by amphibolite facies assemblages (Bloomfield, 1968). However, there are a number of major masses (see Fig. 6) of granulite facies rocks, including silicic, intermediate, and subordinate mafic types, of which the silicic and intermediate types are diopside-hypersthene-quartz-oligoclase (andesine) rocks with K-feldspar, hornblende, and garnet. Hypersthene is partly altered to green amphibole due to retrograde metamorphism, but brown amphibole represents a stable mineral phase, indicating hornblende granulite subfacies conditions (Bloomfield, 1968). Other petrographic types that occur within the granulite facies terrains include marble, calc-silicate rock, sillimanite schist, graphite gneiss, and khondalite composed of sillimanite–K-feldspar–garnet–quartz.

The boundary between granulite facies rocks and the widespread amphibolite facies rocks (see Fig. 6) is gradational, and R. K. Evans (1965) recognized that the amphibolite facies gneiss is the downgraded equivalent of granulite. Similar examples described by Holt (1961), Walshaw (1965), and others demonstrated that the major part of the amphibolite facies terrain of southern Malawi represents inverted granulite facies rocks.

Published radiometric ages for mineral and whole-rock samples from the southern subprovince of Malawi are in the range 400 to 700 m.y. that characterizes the Mozambique Belt (Cahen and Snelling, 1966; Clifford, 1967). A detailed geochronologic Rb-Sr whole-rock study carried out on the metamorphic rocks and associated intrusive and pegmatite rocks (Clifford and others, unpub. data) has demonstrated that (1) the granulite facies metamorphism of southern Malawi is at least 1,100 m.y. old; (2) the superposed amphibolite facies metamorphism took place 718 ± 25 m.y. ago and was accompanied by folding and emplacement of intrusive rocks; (3) a subsequent event of cataclasis (Bloomfield, 1968) represented structural readjustment of a competent crystalline basement; and (4) pegmatite emplacement and the intrusion of posttectonic igneous rocks are dated at 493 ± 16 m.y. (Table 3).

In the northern subprovince of Malawi, the Damaran-Katangan orogenesis has been

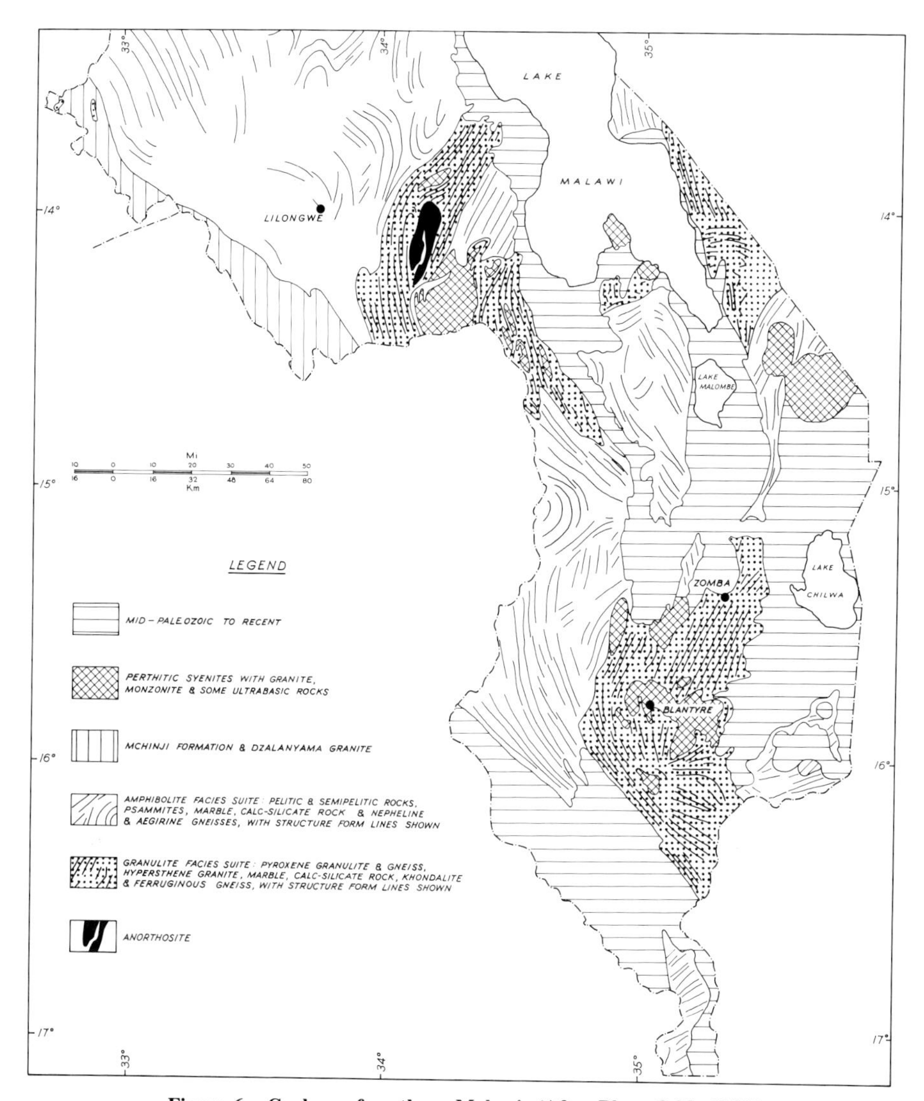

Figure 6. Geology of southern Malawi. (After Bloomfield, 1968.)

imprinted on rocks that had previously been involved in Eburnian-Huabian activity in the Ubendian Belt, and (or) in the Kibaran orogeny in the Irumide Belt (Cannon and others, 1969, p. 2617). The Jembia River Granulite that occurs in this complex junction region has already been discussed earlier in this paper.

Tanzania

A large number of occurrences of granulite facies rocks have been recorded from the Tanzanian segment of the Mozambique Belt (Teale and others, 1935; Temperley, 1938;

Harpum, 1954; Fozzard, 1958; Wright and James, 1958), and Quennell and others (1956) suggested that they form part of a lower member of the Usagaran System. The first published account of extensive granulite-charnockite types in eastern Tanzania was by Stockley (1948). Important masses have since been mapped in the Uluguru Mountains (Sampson and Wright, 1964), the North Pare Mountains (Bagnall, 1963, 1964) and elsewhere (James, 1951, 1958).

Three groups of high-grade metamorphic rocks make up the Usagaran System in the Uluguru Mountains (Sampson and Wright, 1964): (1) a Granulite Group, (2) an Acid (Gneiss) Group, and (3) a Limestone Group (Fig. 7). Of these, the Granulite Group is interpreted in terms of a lower structural unit of garnet-pyroxene granulite and an upper structural unit in which pyroxene gives way to hornblende and biotite and in which garnet-kyanite gneiss and graphite gneiss are common; a large body of garnetiferous meta-anorthosite is conformable with the layering of the granulite (Fig. 7). In contrast, the Acid (Gneiss) Group consists largely of biotite gneiss and hornblende gneiss, containing local pyroxene and the relict textures of granulite. Sampson and Wright (1964) considered that the area represents regional retrogressive metamorphism of original pyroxene granulite to hornblende and biotite granulite. The surrounding regions of the Acid (Gneiss) Group, in the amphibolite facies, are also believed to be the retrograde end-product of this process (Sampson and Wright, 1964, p. 51).

Bagnall (1963, 1964) and Bagnall and others (1965) did a comparable analysis for the North Pare Mountains, where granulite facies assemblages include garnet-quartz-feldspar granulite and enderbite, pyroxene and hornblende granulite, kyanite-sillimanite-garnet-quartzite, graphitic gneiss, metacalcareous rock, amphibolite, metagabbro, and meta-anorthosite; Bagnall (1964) referred to this whole-rock assemblage as the "enderbite-anorthosite layer."

This suite is mantled, on the west and east, by amphibolite facies assemblages that contain some pyroxene granulite (Bagnall, 1963, p. 11) and relict hypersthene in rocks otherwise characterized by hornblende, diopside, garnet, and microcline. Bagnall (1964) referred to this suite as the "migmatite zone" and suggested that it represents an upper crustal series that owes its metamorphic character to the upward diffusion of K, Al, and Si consequent on the formation of the underlying enderbite-anorthosite layer.

Only two whole-rock geochronologic studies of granulite facies suites in the Tanzania segment of the Mozambique Belt have been published. Five pyroxene granulites from the Pare Mountains gave a Rb-Sr isochron age of 936 ± 63 m.y., and samples of granulite and amphibolite from Loibor-Serrit to the east gave an age of 731 ± 8 m.y. (Spooner, 1969; Spooner and others, 1970).

Kenya

The basement within the Mozambique Belt in Kenya (Fig. 4) is composed largely of gneiss, granulite, quartzite, marble, and amphibolite. Except along the western border of the belt, and more locally within it, the rocks are generally in the amphibolite facies. In this terrain, hypersthene-bearing gneiss, granulite, and amphibolite occur locally (Miller, 1956; Dodson, 1963; Matheson, 1966; Walsh, 1966). There are also several discrete bodies of charnockitic rock (Pulfrey, 1946; Schoeman, 1951, 1952; Bear, 1952; Searle, 1952; Sanders, 1954; Williams, 1966), which Schoeman (1951, 1952) subdivided into (1) mafic and ultramafic types, regarded as intrusive, including perknite, websterite, pyroxenite, pyroxene

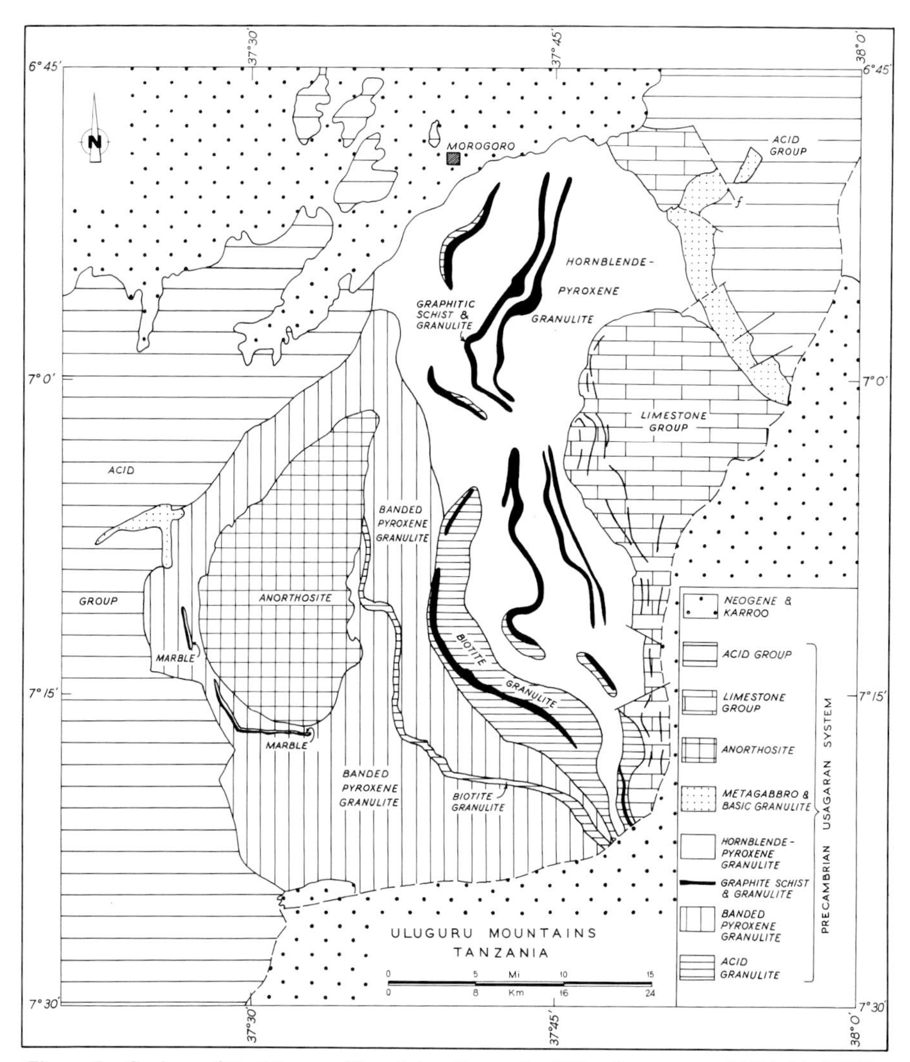

Figure 7. Geology of the Uluguru Mountains, Tanzania. (After Sampson and Wright, 1964.)

hornblendite, norite, hyperite, gabbro, bojite, and anorthosite; and (2) intermediate and felsic and silicic types, including diorite, quartz diorite, and granodiorite. Within these bodies, the predominant mafic and ultramafic rocks are generally concordant with the regional structure of the country rocks and have been metamorphosed in the amphibolite facies. Similar relations are recorded from almost all localities (Bear, 1955; Baker, 1963; Joubert, 1966; Saggerson, 1957; Searle, 1952; Williams, 1966).

The age of these suites in Kenya is unknown. Amphibole from an amphibolite inclusion in charnockite has given a K-Ar age of 855 ± 30 m.y. (Cahen and Snelling, 1966, p. 48) that provides a minimum age for the hypersthene-bearing intrusive rocks.

Uganda

From Kenya, the Mozambique Belt trends westward through northern Uganda into northeastern Zaire (Congo) and northward through Ethiopia into the Sudan. Of these regions, a large number of occurrences of granulite facies rocks are now known in Uganda (see Fig. 8; R. Macdonald, 1966). Early work by Groves (1935) in the northwestern part of Uganda has been a widely quoted standard for African hypersthene-bearing rocks for more than 30 years. In more recent work on Groves's type area of Mount Wati, Macdonald (1964) grouped the granulite facies rocks under the term "Watian" and noted that they are mostly of silicic and intermediate composition. For adjacent areas, Hepworth (1964a, 1964b) recognized that this group also includes kyanite quartzite, cordierite-sillimanite granulite, kyanite-rutile schist, spinel-bearing granulite, and anorthosite. Both Hepworth (1964b, p. 179) and Macdonald (1964) noted that rocks of the Watian suite show evidence of retrograde metamorphism from granulite to amphibolite facies.

Macdonald (1964) suggested that the Watian suites represent lower or infrabasement inliers and that the surrounding gneiss records at least three important phases of post-Watian tectonism: Aruan tectonism accompanied by amphibolite facies metamorphism; Mirian tectonism accompanied by epidote amphibolite facies metamorphism; and Chuan (Madian) tectonism, during which mylonitization affected the region (Hepworth and Macdonald, 1966).

Spooner (1969) and Spooner and others (1970) reported a Rb-Sr isochron age of 2,655 ±

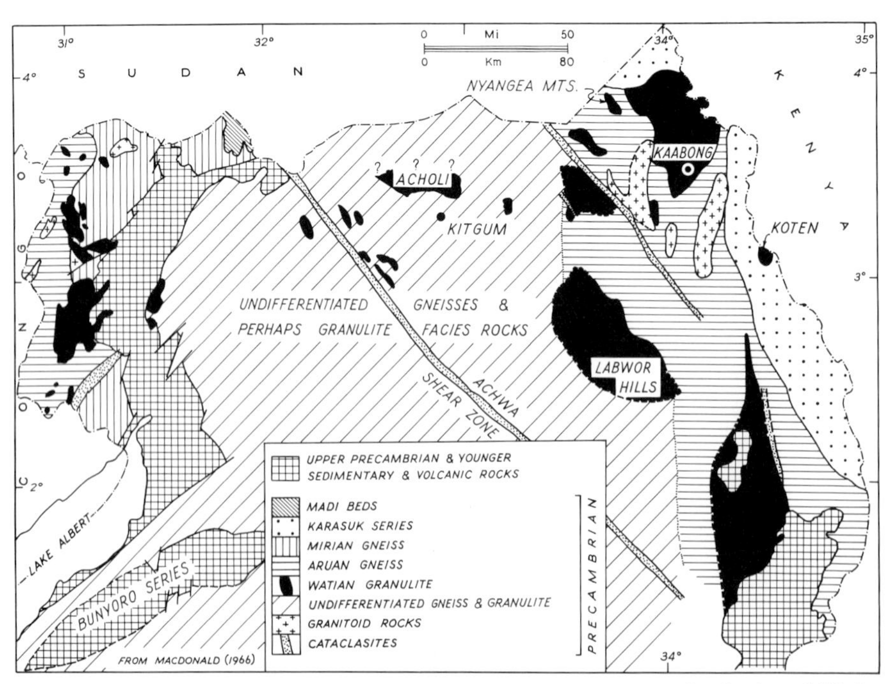

Figure 8. Geology of northern Uganda with particular reference to distribution of Watian granulite facies rocks (in black). (After R. Macdonald, 1966.)

117 m.y. for whole-rock samples from the Watian suite, and Leggo and others (1971) reported discordant U-Pb age patterns that suggest a minimum age of 2,900 m.y. for the initial crystallization of zircon (Table 1). Previous work had yielded K-Ar ages of 660 ± 25 m.y. for biotite from granulite, and 540 ± 20 m.y. for fuchsite from quartzite in the mantling gneiss (Cahen and Snelling, 1966, p. 48); Hepworth and Macdonald (1966) suggested that these ages reflect the Mirian and perhaps Chuan tectonisms.

In addition to this important region in the West Nile, several massifs of granulite facies rocks are known in the central part of northern Uganda (Fig. 8). In particular, pyroxene granulite and gneiss form the Acholi Massif (Almond, 1962); to the south, these rocks are replaced by leucocratic and biotite gneiss in the amphibolite facies, and Almond (1962, p. 7) considered that the granulite facies rock formerly occupied a much more extensive area. Leggo and others (1971) obtained discordant U-Pb ages of as much as 2,500 m.y. for zircon from amphibolite facies gneiss 80 mi (130 km) south of Acholi (see Fig. 8).

Farther to the east, in eastern Uganda, a large number of masses of granulite facies rocks have been recognized (Fig. 8). Of these, the southernmost massif consists of quartz-hypersthene and quartz-clinopyroxene gneiss, sillimanite gneiss, garnet amphibolite, and graphite gneiss (Trendall, 1961, 1965a, 1965b). These suites extend northward to the Sudan border; Hepworth and Macdonald (1966, p. 727) and Hepworth (1967) equated the primary tectonism of all of them with the Watian tectonism of the West Nile (Fig. 8). They include the Kaabong body (Fig. 8; MacGregor, 1962), which is predominantly enderbitic pyroxene gneiss with interbedded biotite-garnet granulite and sillimanite gneiss; retrograde hornblende and biotite occur in the pyroxene gneiss.

The Labwor body (Fig. 8) is particularly interesting: Nixon and others (1973) have described sapphirine-bearing granulites that are believed to represent metamorphosed ferruginous shales with siliceous bands.

Sudan-Ethiopia

Charnockitic rocks are recorded from a number of localities in Sudan (Groves, 1935; Howie, 1958), notably in the south, where they represent the northward extension of the Watian suites of Uganda. Similar rocks were also identified to the south of Khartoum (Howie, 1958).

Granulite facies rocks were described for the first time in the Mozambique Belt in Ethiopia by Kazmin (1972). They form part of the Konso Group, which includes clinopyroxene-hypersthene gneiss, pyroxene amphibolite and interbanded garnetiferous rock, and graphite gneiss, together with bodies of gabbro and pyroxenite. Amphibolization of the pyroxene records retrograde metamorphism that probably took place during the amphibolite facies metamorphism that characterizes the surrounding regions.

Central African Republic

Granulite facies rocks are well represented in the Precambrian basement of the Central African Republic (see Fig. 4; Mestraud, 1964, 1974). The largest occurrence is the Bouca Massif, which occupies an area of 15,000 km^2 (Gérard and Gérard, 1953; Pouit, 1958, 1959; Gérard, 1963; Wacrenier and Wolff, 1965). The most detailed study of this suite is that of Pouit (1958, 1959), who demonstrated that granulite of intermediate composition composes 70 percent of the northeastern segment of the complex and includes pyroxene gneiss and biotite gneiss that contains accessory pyroxene and hornblende. Silicic types constitute 20 percent of the suite and include garnet gneiss (leptynite), augen and banded gneiss, and hypersthene-bearing granitic rocks. Mafic types make up the remaining 10 percent. Several smaller masses of granulite facies rocks occur around Fort Crampel (Pouit, 1959, p. 63; Mestraud, 1964). In Oubangui, Delafosse (1960) identified two similar occurrences, and other examples have been recognized in neighboring regions by Wolff (1963) and Bessoles (1955, 1962).

These various examples indicate granulite facies metamorphism (Pouit, 1959, p. 133 and following), but they occur within a basement of lower metamorphic grade. Pouit (1958, p. 290) suggested that these contrasted suites represent basement and cover, respectively, and that the amphibolite facies metamorphism of the cover has resulted in retrograde metamorphism of the older granulite facies substratum (Pouit, 1958; Mestraud, 1964). Three Rb-Sr ages in the range of 610 to 690 m.y., obtained for biotites from hypersthene-bearing rocks from Oubangui (Roubault and others, 1965), clearly represent minimum ages for the granulite facies metamorphism.

Cameroons-Gabon

In the Cameroons, a number of small bodies of granulite-charnockite suites occur in the Damaran-Katangan (Pan-African) zone of orogenesis (Champetier de Ribes and Reyre, 1959; Dumort, 1968). These are most logically considered as remnants of the Ntem Complex, which lies immediately to the south (see Ancient Nuclei), preserved in the younger zone of tectonism. In western Gabon, Dévigne (1958) recognized the local presence of hypersthene-bearing and diopside-bearing gneiss, migmatite, and diorite forming part of a metamorphic complex of the Pre-Mayombian System, which is believed to be older than 2,500 m.y. (Cahen and Snelling, 1966, p. 135).

Nigeria

The Precambrian geology of Nigeria (Fig. 4) has been described in terms of three major rock suites: (1) an Ancient Gneiss Complex of gneiss and metasedimentary rocks affected by anatexis and migmatization ~2,350 m.y. ago; (2) Low Grade Metasediments deposited more than ~700 m.y. ago, now preserved as narrow, linear belts superposed on the older basement; and (3) the Older Granite plutons intruded ~635 m.y. ago (Grant, 1970; N. K. Grant and J. L. Powell, 1971, unpub. data). Within (1) and (3) are examples of hypersthene-bearing rocks (Fig. 9). These include bodies of dioritic, gabbroic, and pyroxenitic rocks that are structurally conformable with the Ancient Gneiss Complex in southwestern Nigeria (Jones and Hockey, 1964, p. 43–44). The diorite and gabbro are characterized by the coexistence of hypersthene and augite and by partial replacement of pyroxene by hornblende. Jones and Hockey (1964) recognized (1) that these hypersthene-bearing rocks are intrusive into the Ancient Gneiss Complex, but their emplacement predated the granite-gneiss event that has yielded a Rb-Sr isochron age of 2,340 ± 70 m.y. (Grant, 1970); and (2) that they have the relict mineralogic characteristics of the granulite facies, whereas the enclosing gneiss has amphibolite facies assemblages.

A petrologically similar suite has been identified by Carter and others (1963) in northeastern Nigeria, where numerous examples occur as xenoliths in the Older Granite plutons. In addition, Oyawoye (1961, 1964) described orthopyroxene-bearing quartz diorite, and he considered that this rock type and the associated fayalite-quartz monzonite (bauchite) are related to charnockite. To the west of these occurrences, Truswell and Cope (1963) recognized hypersthene-bearing dioritic rocks. Most of these bodies are apparently related to the ~640-m.y.-old Older Granites (Truswell and Cope, 1963) and therefore are considerably younger than those of southwestern Nigeria.

Dahomey-Togo

This region sits astride the western margin of the Damaran-Katangan (Pan-African) orogenic zone that extends through Dahomey, Togo, Ghana, Nigeria, and Mali and northward into the Hoggar (Figs. 1, 4). Within this zone, the crystalline rocks of Nigeria extend westward as the Dahomeyan System (Roques, 1948). In this complex assemblage, hypersthene-bearing rocks occur as part of two distinctive groups: (1) intrusive granitic orthogneiss (Kouandé-Anié Group), for which Bonhomme (1962) obtained a single whole-rock Rb-Sr age of 1,750 ± 230 m.y.; and (2) a group of mafic and ultramafic rocks (Kabré-Dérouvarou Group; Pougnet, 1957; Aicard, 1957). Of these, the latter are widely scattered and consist of pyroxene gneiss, pyroxenite, amphibole gneiss, amphibolite, and eclogite (Aicard, 1957, p. 76). These essentially gabbroic masses are regarded as intrusive rocks (Pougnet, 1957) metamorphosed in the granulite facies (Aicard, 1957) and subsequently subjected, particularly in their margins, to the amphibolite facies metamorphism.

The petrologic similarity between these Kabré-Dérouvarou bodies and the largely mafic charnockitic suites of Kenya is striking. Equally interesting is the distribution of these masses (Fig. 9), which seem to be concentrated in a well-defined north-northeast–trending zone that

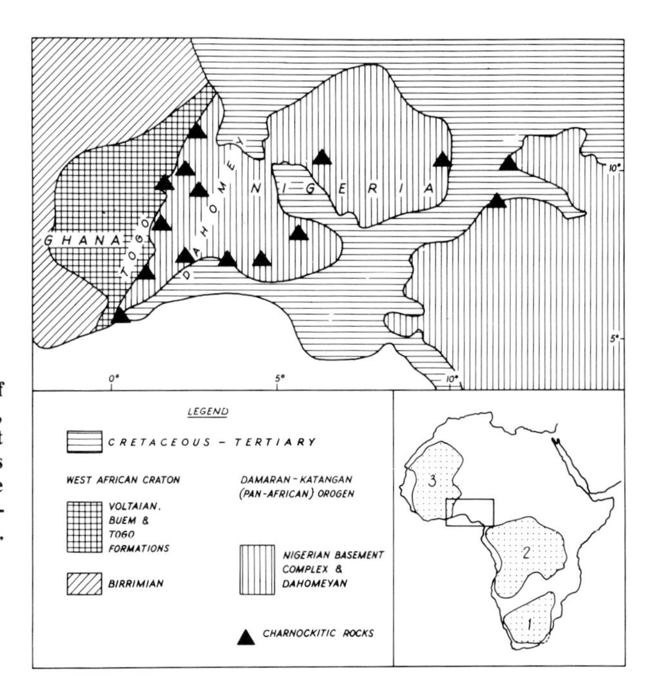

Figure 9. Distribution of charnockitic, largely basic, rocks in West Africa. Inset shows stable cratonic blocks (1, 2, and 3) during late Precambrian–early Paleozoic time (Clifford, 1970).

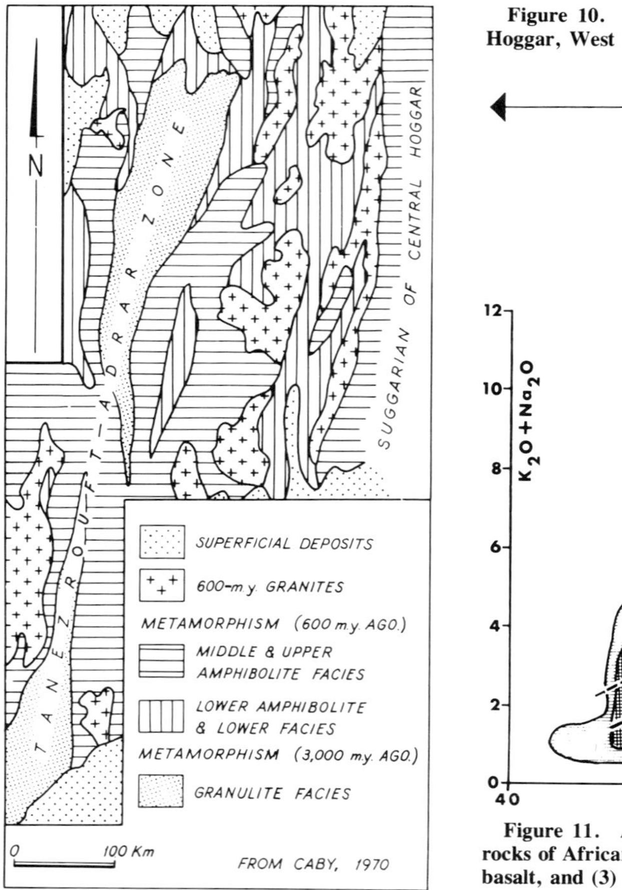

Figure 10. Distribution of metamorphic facies in vicinity of Tanezrouft-Adrar zone of the Hoggar, West Africa. (After Caby, 1970.)

K_2O+Na_2O

SiO_2

Contours at 1, 2, 3 & 5 %

Figure 11. Alkali-silica diagram showing contoured distribution of 240 hypersthene-bearing rocks of African charnockite-granulite suites in relation to fields of (1) tholeiite, (2) high-alumina basalt, and (3) alkali olivine basalt. (Given by Kuno, 1968, p. 627.)

extends from southeastern Ghana through Togo and into Dahomey, parallel with and close to the western margin of the Damaran-Katangan zone of orogeny. Finally, the emplacement of these mafic rocks predated a major event of migmatization (the Pira-Chra Group) that is most logically correlated with the 2,350 m.y. event recorded in southwestern Nigeria (Grant, 1970).

Ghana

The Dahomeyan System of Togo and Dahomey extends southwestward into Ghana (Fig. 9; Junner, 1940; Junner and Bates, 1945; Bates, 1957). In this region, there is a north-northeast–trending zone composed of a variety of granulite facies types (including garnet-bearing hornblende and pyroxene gneiss and granulite) and hypersthene-augite pyroxenite (Junner, 1940). Eclogite is present, as is garnet-hornblende-pyroxene-scapolite gneiss with marked eclogitic affinities (Von Knorring and Kennedy, 1958).

Hoggar

The central Sahara is underlain by Precambrian rocks that have traditionally been subdivided into the Suggarian and Pharusian (Rocci, 1965; Black and Girod, 1970). Of these, the Suggarian is a suite of higher grade metamorphic rocks forming three north-trending zones that are, from west to east, the Tanezrouft-Adrar zone, the Central Hoggar zone and the Anahef-Air zone (Lelubre, 1952a, 1952b, 1969). Of particular importance here is the presence of hypersthene-bearing crystalline rocks in the Tanezrouft-Adrar zone (Lelubre, 1952a, p. 57); in addition, small bodies have been identified in the northeastern part of the Central Hoggar zone (Lelubre, 1952a; Ferragne, 1964; Bertrand, 1967) and elsewhere (Reboul and others, 1962, p. 19). These rocks were originally regarded as a facies of the Suggarian, but in recent years, it has been suggested that they represent an older suite, for which Lelubre (1969, p. 28–29) has proposed the term "Ouzzalian."

The Ouzzalian suite forms two masses in the Tanezrouft-Adrar zone (Fig. 10), and studies by Lelubre (1952a), Radier (1959), Karpoff (1958), Giraud (1961, 1964), and Le Fur (1966) have shown that it includes silicic and mafic granulite; leptynite; hypersthene-magnetite quartzite; calc-silicate rock and marble; garnet, sillimanite, cordierite, and graphite gneiss; pyroxene amphibolite; and a number of igneous rock types ranging from pyroxenite, norite, and gabbro to granite and syenite. This huge mass of granulite facies metamorphic rocks and associated intrusive charnockitic types lies within the Damaran-Katangan (Pan-African) orogenic zone (Kennedy, 1965; Clifford, 1968, p. 309; Caby, 1970) as evidenced by (1) U-Pb zircon ages of 650 m.y. for syntectonic granite in both the Pharusian and Suggarian terrains, and mica and whole-rock Rb-Sr and K-Ar ages of 450 to 700 m.y. (Eberhardt and others, 1963; Lay and Ledent, 1963; Lay and others, 1965; Picciotto and others, 1965); and (2) whole-rock Rb-Sr isochron ages of 560 ± 40 m.y. and mica ages of 500 to 525 m.y. for late tectonic granites (Boissonnas and others, 1969). Nevertheless, Ferrara and Gravelle (1966) obtained a whole-rock Rb-Sr isochron age of 3,030 m.y. for samples of alkali granite, calc-alkali syenite, granitoid gneiss, and hypersthene-bearing granitic gneiss from the Ouzzalian suite.

Sierra Leone

The coastal area of Sierra Leone (Figs. 1, 4) represents a segment of the Damaran-Katangan (Pan-African) orogenic zone. Within this zone, high-grade gneiss (the Kasila Gneiss) forms a belt about 30 km wide parallel to the coast, and consists of a variety of metamorphic rocks, including hypersthene gneiss and granulite (Pollett, 1951). Most of the mineral and whole-rock samples from the Kasila Gneiss have yielded ages of 480 to 650 m.y. (Allen and others, 1967; Hurley and others, 1971), but Hurley and others (1971, p. 3487) obtained an apparent whole-rock Rb-Sr age of ~2,600 m.y. for quartz-plagioclase-hypersthene-garnet-biotite granulite; pyroxene from garnet-pyroxene granulite has yielded a K-Ar age of 1,340 ± 130 m.y. (Allen and others, 1967).

HERCYNIAN DOMAIN

The limited zones of middle and upper Paleozoic orogenesis in Africa (Fig. 1) exhibit few examples of high-grade metamorphic rocks; the principal exception is the Mauritanide zone of northwestern Africa (Sougy, 1969). Within the northern sector of this zone, Precambrian rocks emerge from beneath the deformed lower Paleozoic rocks in a number of inliers and thrust slices. Among the Precambrian rocks, Arribas (1960) has identified a group of rocks of charnockitic character, including hypersthene and amphibole gneiss, pyroxene quartzite, garnet gneiss, noritic gabbro, and pyroxenite. This suite forms part of the Auhaifrit Series, which has been correlated with the Amsaga Group of Mauritania (more fully described in the section dealing with the Eburnian-Huabian domain).

DISCUSSION

Two well-defined events of granulite facies metamorphism and intrusion of charnockitic rocks at ~2,900 to 3,000 and 1,200 m.y. ago are recognized at the present erosion level in Africa; comparable events at ~2,900 and 1,000 m.y. ago have been recognized in Madagascar (Vachette and others, 1969). Of these, evidence for the 1,200-m.y. event is most extensively preserved in the Namaqualand-Natal segment of the Kibaran orogenic zone. In contrast, examples of the ~3,000-m.y. granulite-charnockite suites occur widely in Africa — for example, in the Kasai, Sierra Leone–Ivory Coast, and Gabon-Cameroons nuclei. In addition, several suites found as remnants in more modern orogenic zones are logically included with the 3,000-m.y. event: (1) in Eburnian domains in the Limpopo Belt and the Ubendian Belt, and in Mauritania, Angola, and Gabon; (2) in Damaran-Katangan domains in the Hoggar, Uganda, Nigeria, Sierra Leone, Rhodesia, and perhaps Mozambique; and (3) in Hercynian domains in Spanish Sahara (see Table 1).

As elsewhere in the world, the African granulite-charnockite suites are petrographically heterogeneous. Among the constituent rock types, it is generally agreed that quartzite, highly aluminous gneiss and schist, ironstone, graphite gneiss, and marble are of metasedimentary origin. In contrast, the origin of the typomorphic hypersthene-bearing granulite and gneiss and of certain supposed intrusive rocks of charnockitic type has been the subject of debate by African authors. Contoured plots of $(K_2O + Na_2O)/SiO_2$, and of $A(K_2O + Na_2O)$, F (total iron as FeO), and M(MgO) for 240 African hypersthene-bearing granu-

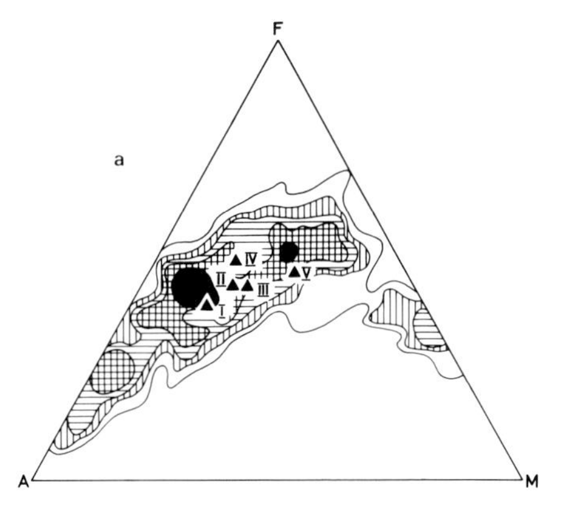

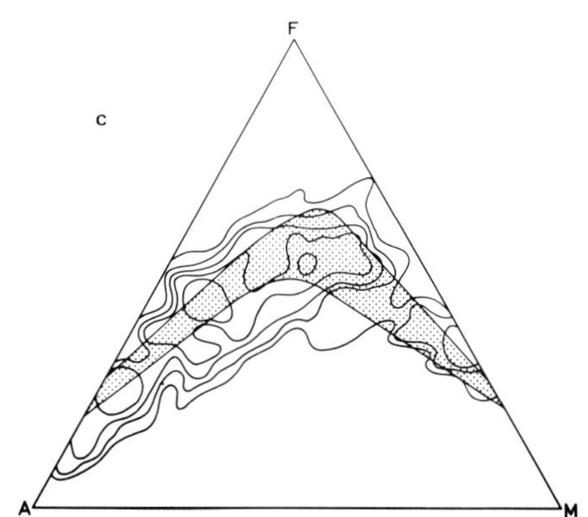

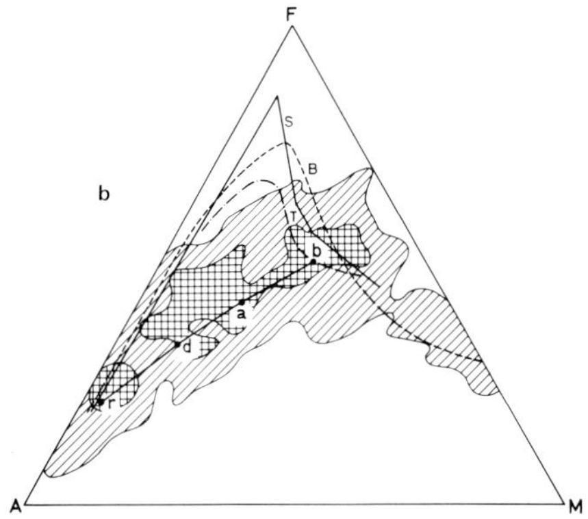

Figure 12. a. AFM diagram for 240 hypersthene-bearing rocks of African charnockite-granulite suites, contoured at 1, 2, 3, 4, and 6 percent. Weighted average compositions of individual African suites shown as follows: II, Uluguru Mountains, Tanzania; III, Lulua Massif, Kasai; and IV, Bouca, Central African Republic (see Table 4). Unweighted mean compositions for suites shown as follows: I, southern Malawi; V, Katumbi, Tanzania. b. Data from Figure 12a contoured at 1 and 4 percent, shown in relation to Daly's (1933) average rhyolite (r), dacite (d), andesite (a), and basalt (b), and to fractionation trends in B, Bushveld Igneous Complex (Liebenberg, 1969); S, Skaergaard intrusion (Macdonald and Katsura, 1964); and T, Tasmanian dolerites (Macdonald and Katsura, 1964). c. Data from Figure 12a shown in relation to field (stippled) of Hawaiian tholeiitic and alkalic rocks. (After G. A. Macdonald, 1968.)

lites and charnockitic rocks are given in Figures 11 and 12a, respectively. Figure 11 shows the clear correlation between (1) the field of maximum concentration of granulite-charnockite suites; and (2) the field of high-alumina basalt and its fractionation products, notably its calc-alkalic hypersthenic rock series derivatives (Kuno, 1968, p. 682–683). Equally, there is a close relation between the AFM trend for the African hypersthenic rocks and (1) the trend of Daly's average basalt-andesite-dacite-rhyolite calc-alkalic trend (Fig. 12b) and (2) the Hawaiian tholeiitic and alkalic trend (Fig. 12c).

Chemical correlations of the type illustrated in Figures 11 and 12 must be interpreted with some caution, particularly in light of the views of Lambert and Heier (1968, p. 43) that granulite facies metamorphism results in the upward migration of K, Si, Rb, Pb, Th, and U into regions of lower pressure, resulting in lower abundances of these elements and higher abundances of Ca, Fe, Mg, Mn, and Ti in deeper regions of medium to high pressure. These two pressure regimes are based on Green and Ringwood's (1967) subdivision of low-pressure and medium- to high-pressure granulites with the following typomorphic minerals:

(1) low-pressure granulite, characterized by olivine + plagioclase and by the occurrence of cordierite and andalusite; (2) medium-pressure granulite, characterized by the association plagioclase + hypersthene ± clinopyroxene ± quartz; olivine + plagioclase is unstable, and the typical aluminosilicates are kyanite and sillimanite; and (3) high-pressure granulite, containing garnet + clinopyroxene + hypersthene + plagioclase (see also Lambert and Heier, 1968, p. 32). To date, olivine + plagioclase has been identified only in a limited number of suites in Natal, Kenya, and Nigeria. Almost all other African examples contain plagioclase + hypersthene + clinopyroxene ± garnet ± quartz in rocks of appropriate composition. It is, therefore, of interest to note that although average compositions are available only for three examples of African granulite-charnockite suites (Table 4, I through III), they are comparable to those of medium- to high-pressure granulites elsewhere in the world (Table 4, B, C, D) and contrast with the average composition of the low-pressure granulites of Cape Naturaliste (Table 4, A).

Heier (1964) and Lambert and Heier (1968) suggested that granulite facies metamorphism results in the development of a deeper crustal zone of medium- to high-pressure metamorphism, depleted in Rb, with concomitant small growth of Sr^{87} with time; and a shallower zone enriched in Sr^{87} and Rb. Table 5 lists the ages and initial Sr^{87}/Sr^{86} ratios (R_o) for hypersthene-bearing granulite-charnockite suites of Africa (nos. 1 through 10) and elsewhere in the world (nos. 12 through 29), and these data are plotted in Figure 13a. Several models have been proposed for the Sr-isotopic evolution in the mantle and crust (Faure and Hurley, 1963; Steuber and Murthy, 1966; Davies and others, 1970). All assume that the Earth formed 4,500 m.y. ago with a uniform R_o of 0.699 and that since that time, Sr^{87} has grown in the mantle (line AB of Figure 13a; Davies and others, 1970, p. 587–588). Figure 13a shows that a number of granulite-charnockite suites (Table 5, nos. 3, 5, 20, 21, 27, 29)

TABLE 4. AVERAGE COMPOSITIONS OF SOME AFRICAN AND WORLD GRANULITE SUITES

	I	II	III	A	B	C	D	E
SiO_2	61.0	60.6	58.8	67.6	61.2	60.7	60.6	54.8
TiO_2	1.0	0.6	1.3	0.7	0.6	0.9	0.9	1.3
Al_2O_3	16.1	17.8	15.9	13.6	16.4	15.8	15.4	15.5
Fe_2O_3	1.7	1.5	2.8	4.5*	2.3	1.7	7.2*	11.3*
FeO	5.0	4.3	5.4	. .	3.5	5.5	. .	. .
MgO	2.8	2.9	2.7	1.3	3.0	3.0	3.9	4.4
CaO	6.1	7.3	6.9	2.9	4.4	5.3	5.7	7.4
Na_2O	3.8	3.4	3.4	3.0	4.0	3.5	2.8	2.1
K_2O	1.8	1.1	1.9	4.5	3.0	2.3	2.6	1.7

Column:
I, Uluguru Mountains, Tanzania.
II, Lulua Massif, Kasai, Zaire (Congo).
III, Bouca Massif, Central African Republic.
A, Cape Naturaliste, Australia (Lambert and Heier, 1968).
B, Northern Norway (Heier and Thoresen, 1971).
C, Brazilian Shield (Sighinolfi, 1971).
D, Musgrave Range, Australia (Lambert and Heier, 1968).
E, Fraser Range, Australia (Lambert and Heier, 1968).

* Total iron as Fe_2O_3.

TABLE 5. AGES AND Sr^{87}/Sr^{86} INITIAL RATIOS (R_0) FOR CHARNOCKITE-GRANULITE AND ASSOCIATED SUITES IN AFRICA AND ELSEWHERE

No.*	Locality	Age (m.y.)	R_0	Reference
1.	Kenema Assemblage, Sierra Leone	3,020 ± 50	0.7027	Andrews-Jones, 1968
2.	Hoggar, Algeria	3,030	0.709	Ferrara and Gravelle, 1966
3.	Lulua Massif, Kasai	3,020 ± 125	0.6976 ± 0.0054	Delhal and Ledent, 1971
4.	Man Massif, Ivory Coast	2,915 ± 115	0.707 ± 0.001	Papon and others, 1968
5.	Man Massif, Ivory Coast	2,860 ± 140	0.699 ± 0.001	Papon and others, 1968
6.	West Nile, Uganda	2,655 ± 117	0.7054 ± 0.001	Spooner and others, 1970
7.	Southern Malawi (granulites)	1,126 ± 29	0.7055 ± 0.0014	Clifford and others, in prep.
8.	Nababeep, Namaqualand	1,213 ± 22	0.7191 ± 0.0021	Clifford and others, 1974
9.	Pare Mountains, Tanzania	936 ± 63	0.7056 ± 0.0011	Spooner and others, 1970
10.	Loibor Serrit, Tanzania	731 ± 8	0.7064 ± 0.0001	Spooner and others, 1970
11.	Southern Malawi (amphibolites; see no. 7 and text)	718 ± 25	0.7063 ± 0.0004	Clifford and others, in prep.
12.	Lewisian, Scotland	2,600	0.7065	C. R. Evans, 1965
13.	Kushalnagar, India	2,618 ± 46	0.7039 ± 0.0005	Spooner and Fairbairn, 1970
14.	Niligiri Hills, India	2,615 ± 80	0.7023 ± 0.0012	Crawford, 1969
15.	Madras, India	2,580 ± 95	0.7059 ± 0.0042	Crawford, 1969
16.	Salem, Madras State, India	2,476 ± 115	0.7042 ± 0.0002	Spooner and Fairbairn, 1970
17.	Pallavaram, Madras State, India	1,980 ± 124	0.7037 ± 0.0007	Spooner, 1969; Spooner and Fairbairn, 1970
18.	Kanuku Complex, Guyana	2,182 ± 95	0.7018 ± 0.0011	Spooner and Fairbairn, 1970
19.	Westport, Ontario, Canada	1,338 ± 47	0.7057 ± 0.0009	Spooner, 1969; Spooner and Fairbairn, 1970
20.	Crane Mountain, New York, U.S.A.	1,336 ± 71	0.7025 ± 0.0025	Spooner and Fairbairn, 1970
21.	Indian Lake, Blue Mountain Lake, and western Canada lakes	1,465 ± 85	0.7014 ± 0.0013	Spooner and Fairbairn, 1970
22.	Fraser Range, Australia	1,328 ± 12	0.7049 ± 0.0012	Arriens and Lambert, 1969
23.	Musgrave Range, Australia (gneisses)	1,380 ± 120	0.7072 ± 0.0025	Arriens and Lambert, 1969
24.	Musgrave Range (Ernabella pyroxene granite)	1,120 ± 100	0.7106 ± 0.0014	Arriens and Lambert, 1969
25.	Highland Series, Ceylon	2,300 ± 20	0.7038 ± 0.002	Wickremasinghe, 1969
26.	Saxony	473	0.712	Jäger and Watznauer, 1969
27.	Langöy, Norway	2,820 ± 50	0.702 ± 0.002	Heier and Compston, 1969
28.	Lofoten, Norway	1,775 ± 30	0.7037 ± 0.0003	Heier and Compston, 1969
29.	Soavinandriana-Betafo, Madagascar†	1,010 ± 108	0.7022 ± 0.0004	Vachette and others, 1969

* Numbers refer to points in Figure 13.

† The charnockitic massif of Manakambahiny-est has also been studied (Vachette and others, 1969). It has yielded an age of 2,920 m.y. and a R_0 value of 0.7055 ± 0.0005; however, because of the error (>500 m.y.) in the age, these data are not plotted in Figure 13.

may directly reflect the Sr-isotopic composition of the mantle. However, most of the suites define a crude path IJ (Fig. 13b). Because these are assemblages bearing two pyroxenes + plagioclase ± garnet in rocks of appropriate composition, path IJ may represent the Sr-isotopic evolution path for the deeper parts of the Earth's crust; on the model of Heier (1964) and Lambert and Heier (1968), it must have been complemented by another path (or paths), of steeper slope, representing the granitophile extract enriched in Rb and Sr^{87}. This interpretation has the merit that the R_o values of anorthosite (Fig. 13b) are consistent with a deeper crustal environment of formation, as has been suggested by Green (1969).

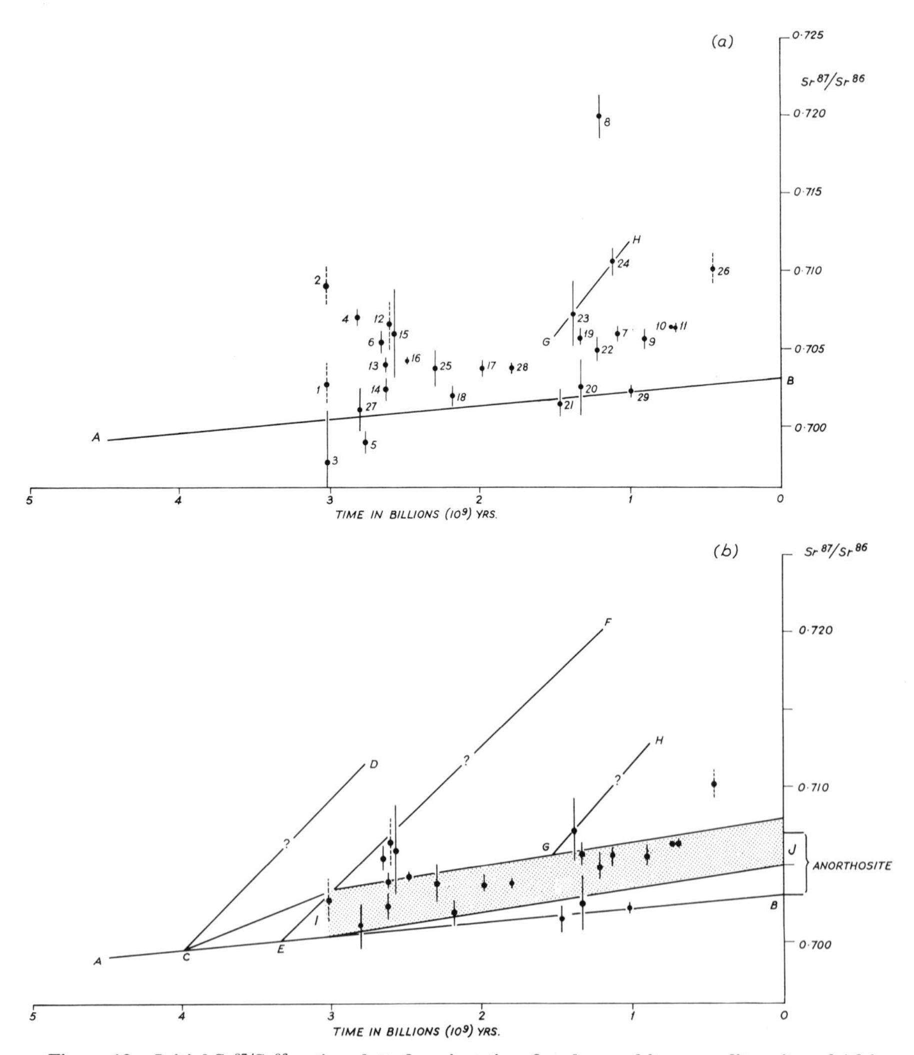

Figure 13. Initial Sr^{87}/Sr^{86} ratios plotted against time for charnockite-granulite suites of Africa and elsewhere. a. Numbered points refer to localities given in Table 5; error bars reflect confidence limit of initial ratios. b. Possible Sr-isotopic evolution paths discussed in text.

In contrast, however, granulite facies metamorphism may be isochemical with respect to the Sr-isotopic composition of the rocks. This interpretation has important implications in the light of the suggestion by Arriens and Lambert (1969) that granulites of the Fraser and Musgrove Ranges of Australia (Fig. 13a, nos. 22, 23) are metamorphosed mantle derivatives; path IJ may then define the Sr-isotopic development in the mantle beneath the continental crust (see Davies and others, 1970, p. 588). On such an inheritance model, granulite suites from the Hoggar (Table 5, no. 2) and Namaqualand (Table 5, no. 8) can be explained in terms of Sr evolution in former crustal materials, perhaps of types CD and EF (Fig. 13b), respectively, prior to granulite facies metamorphism. Moveover, isochemical metamorphism logically explains differing R_o values in coexisting rock suites, such as have been described by Papon and others (1968; see Table 5, nos. 4, 5) and in the elegant work by Gray (1971). No attempt is made here to distinguish this type of interpretation from the allochemical hypothesis of Heier (1964) and Lambert and Heier (1968); it is hoped that the wide range of African granulite-charnockite suites documented in this paper will contribute significantly to the solution of this problem.

Although apparently contradictory in the light of the work of Green and Ringwood (1967), cordierite-bearing rocks occur with medium- to high-pressure granulite notably in the Hoggar, Namaqualand, Uganda, Mauritania, Rhodesia, and Malawi. There is now considerable experimental data on the *P-T* range of cordierite; its development is influenced by the K_2O, CaO, MnO, and TiO_2 content and the P_{H_2O} of the system (Seifert, 1970; Hensen and Green, 1970). Moreover, Figure 14a illustrates that the *maximum* range of pressure of cordierite ± garnet assemblages is rigidly controlled by the $Fe^{2+}/(Fe^{2+} + Mg)$ of the system and that these assemblages may occur with medium- to high-pressure granulites, provided that the $Fe^{2+}/(Fe^{2+} + Mg)$ ratio of the cordierite system is <0.80 to 0.90. This condition is satisfied in African and Madagascan examples of cordierite gneiss and granulite, because they have $Fe^{2+}/(Fe^{2+} + Mg)$ ratios in the general range of 0.30 to 0.70 (Table 6).

TABLE 6. $Fe^{2+}/(Fe^{2+} + Mg)$ RATIOS OF CORDIERITE-BEARING ROCKS, AFRICA AND MADAGASCAR

Rock*	Locality	$\frac{Fe^{2+}}{Fe^{2+}+Mg}$	K_2O (wt %)
a. Cordierite-sillimanite-garnet-biotite gneiss	Mauritania	0.68	3.80
b. Garnet-sillimanite-cordierite granulite	Namaqualand	0.67	0.01
c. Garnet-sillimanite-cordierite-spinel granulite	Namaqualand	0.64	2.36
d. Cordierite-spinel-garnet gneiss	Madagascar	0.64	3.26
e. Cordierite syenitic gneiss	Madagascar	0.58	3.36
f. Cordierite-spinel-garnet gneiss	Madagascar	0.57	2.74
g. Cordierite syenitic gneiss	Madagascar	0.55	2.88
h. Cordierite-sillimanite-garnet gneiss	Mauritania	0.54	0.85
i. Cordierite-biotite-garnet gneiss	Madagascar	0.51	1.81
j. Cordierite-garnet leptynite	Madagascar	0.40	5.90
k. Cordierite-spinel-hypersthene granulite	Namaqualand	0.34	4.14
l. Hypersthene cordierite	Madagascar	0.32	1.92
m. Cordierite-sapphirine rock	Namaqualand	0.11	0.85
n. Enstatite-sillimanite-cordierite rock	Rhodesia	0.02	0.00

* Sources of analytical data: Blanchot (1955), a and h; Chinner and Sweatman (1968), n; Clifford and others (in prep.), m; Noizet (1969), d, e, f, i, j, and l; Von Backström (1964), b, c, and k.

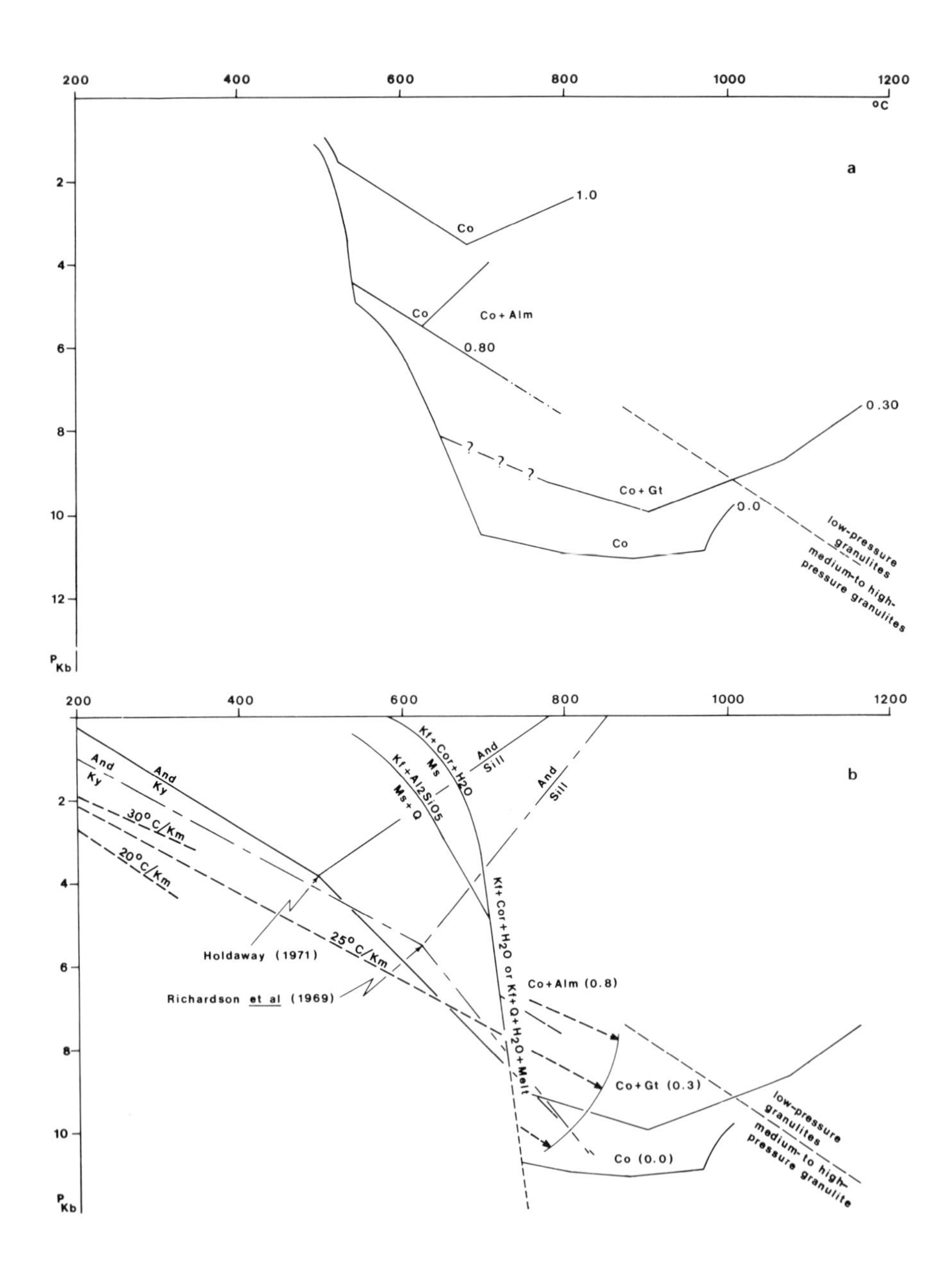

Figure 14. a. Experimental data for stability fields of cordierite ± garnet for systems with $Fe^{2+}/(Fe^{2+} + Mg)$ ratios: 1.0 (Richardson, 1968); 0.80 (Hirschberg and Winkler, 1968); 0.30 (Hensen and Green, 1970); 0.0 (Schreyer and Yoder, 1964). Extrapolated boundary between low-pressure and medium- to high-pressure granulites is from Green and Ringwood (1967, p. 807). b. Cordierite ± garnet data from 14a, shown in relation to experimental data for muscovite and muscovite + quartz (Velde, 1966; Althaus and others, 1970) and the *P-T* fields of andalusite, sillimanite, and kyanite. (After Richardson and others, 1969; Holdaway, 1971.)

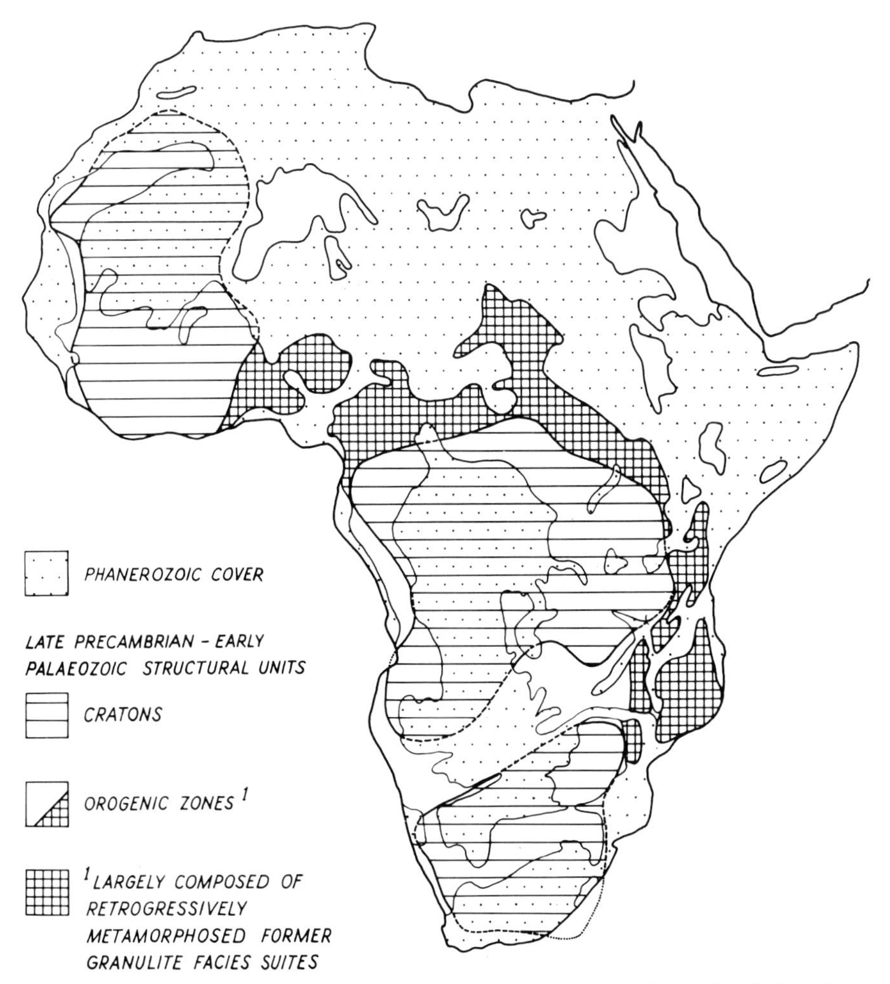

Figure 15. Idealized distribution of retrogressively metamorphosed granulite facies suites, at present erosion level, in late Precambrian–early Paleozoic Damaran-Katangan zone of orogenesis. Remnants of upper Precambrian geosynclinal rocks and other supracrustal sequences have been omitted.

In Figure 14b the data from Figure 14a are shown in relation to the inversion curves of muscovite and muscovite + quartz (Velde, 1966; Althaus and others, 1970) and to the stability fields of kyanite, sillimanite, and andalusite (Richardson and others, 1969; Holdaway, 1971). On the basis of these data, the development of cordierite-bearing assemblages in African granulite suites is consistent with elevated pressures and prograde metamorphic paths representing geothermal gradients *as low as* 25 to 35°C/km;[2] a gradient of that order is

[2] That even lower gradients were operative is clear from the Rhodesian example of cordierite-enstatite-sillimanite rock (Table 6) that records *P-T* conditions of 10 kb or more at 800°C (Chinner and Sweatman, 1968) and implies a crust at least 35 km thick 2,700 to 3,000 m.y. ago (see discussion of the Limpopo Belt).

well illustrated in the prograde metamorphism, via kyanite at lower grades, in Namaqualand (Joubert, 1971, p. 116–122). However, with this exception and others in Sierra Leone, the Ubendian Belt, and the Limpopo Belt, the paths of prograde metamorphism have largely been obliterated by later retrograde metamorphism. This is particularly well illustrated in the Damaran-Katangan orogenic zone, in which the majority of granulite facies suites are significantly older than the ~600-m.y. tectonism and metamorphism (Table 1). The wide scatter of these granulite remnants (Fig. 4) and the extent of their inverted amphibolite facies equivalents suggest that, with the exception of some masses of younger sedimentary and volcanic rocks, a major part of the Mozambique Belt and of its extension through Uganda, Central African Republic, the Cameroons, Ghana, Dahomey, and Nigeria represents a former, much more extensive, zone of antique granulite facies rocks, remnants of which are now preserved as small islands in a sea of amphibolite facies rocks and anatectic migmatites at the present erosion level (Fig. 15). The source of H_2O for this widespread retrogression is not known. Equally intriguing is the interpretation of the physicochemical significance of the gradational interface between the older ''dry'' granulite facies and younger ''wet'' amphibolite facies. One tentatively favored alternative is that this contact represents the upper *P-T* limit of the *new* amphibolite facies metamorphism. If this line of reasoning is correct, it may be logically argued that the granulite remnants were resubjected to granulite facies conditions during the new metamorphism, but in the absence of a pervasive fluid phase, their isotopic clocks were not completely reset at that time. This speculation is currently being tested in an intensive petrochemical, mineral chemical, and isotopic study of the granulite and amphibolite facies suites of southern Malawi (see Fig. 6; Clifford and others, in prep.).

In the foregoing discussion, a number of problems of African granulite-charnockite suites have been raised; many remain unsolved, however, largely because rock and mineral chemical data and isotopic information are not yet sufficiently refined for such a task. This paper is, therefore, presented in a spirit of challenge for future studies; those studies will certainly contribute significantly to the debate on the temporal and spatial evolution of the African crust, particularly its third dimension.

Acknowledgments

During the course of this work, I have benefited from discussions with colleagues in many parts of the world: H. L. Allsopp, D. A. Andrews-Jones, A.R.J. Araujo, P. S. Bagnall, K. P. Bahnemann, J. Barrère, H. Besairie, A. Blanchot, K. Bloomfield, D. Bridgewater, R. Cannon, R. D. Davies, W. S. Fyfe, N. K. Grant, S. Hausmann, J. V. Hepworth, G. Hottin, P. Joubert, J. L. Mestraud, J. R. McIver, P.R.J. Naidu, L. O. Nicolaysen, G. Noizet, A. Razafiniparany, O. C. Wickremasinghe, A. F. Wilson, B. F. Windley, and H.G.F. Winkler.

The early part of this study was carried out under the aegis of the Research Institute of African Geology, which was supported through the generosity of the Anglo-American Corporation Ltd. The later stages of the work were financed by a research grant from the University of Witwatersrand.

I am grateful to the individuals and organizations mentioned above and to S. M. McCarthy and L. Hansen for secretarial and clerical services, M. Hudson for painstaking photographic work, M. Miles for drafting Figure 11, and to L. Clifford for preparing Figure 14.

Finally, it is appropriate to record my gratitude to D. M. Morgan and others who acted as reviewers for this paper. Their comments resulted in improvements in the presentation of this work.

References Cited

Aicard, P., 1957, Le Précambrien du Togo et du nord-ouest du Dahomey: French West Africa, Dir. Fédérale Mines et Géologie Bull., v. 23, 226 p.

Allen, P. M., Snelling, N. J., and Rex, D. C., 1967, Age determinations from Sierra Leone: Massachusetts Inst. Technology, Dept. Geology and Geophysics, 15th Ann. Prog. Rept., p. 17–22.

Allsopp, H. L., Davies, R. D., de Gasparis, A.A.A., and Nicolaysen, L. O., 1969, Review of Rb-Sr age measurements from the early Precambrian terrain in the south-eastern Transvaal and Swaziland: Geol. Soc. South Africa Spec. Pub. 2, p. 433–444.

Almond, D. C., 1962, Explanation of the geology of sheet 15 (Kitgum): Uganda Geol. Survey Rept., v. 8, 55 p.

Althaus, E., Karotke, E., Nitsch, K. H., and Winkler, H.G.F., 1970, An experimental re-examination of the upper stability limit of muscovite plus quartz: Neues Jahrb. Mineralogie Monatsh., v. 7, p. 325–336.

Andrews-Jones, D. A., 1968, Petrogenesis and geochemistry of the rocks of the Kenema District, Sierra Leone [Ph.D. thesis]: Leeds, England, Univ. Leeds, 171 p.

Araujo, J.R.F., 1966, The Mozambique Belt in the Barué area, Manica e Sofala District, Mozambique, with special reference to the petrology, stratigraphy and metamorphism [Ph.D. thesis]: Leeds, England, Univ. Leeds, 166 p.

——1967, The Mozambique Belt in the Barué area, Manica e Sofala District, Mozambique, with special reference to the petrology, stratigraphy and metamorphism: Univ. Leeds, Research Inst. African Geology, 11th Ann. Rept., p. 19.

Arribas, A., 1960, Las formaciones metamorficas de Sahara Español y sus relaciones con el Precambrico de otras regiones africanos: Internat. Geol. Cong., 21st, Copenhagen 1960, Comptes Rendus, pt. 9, p. 51–58.

Arriens, P. A., and Lambert, I. B., 1969, On the age and strontium isotopic geochemistry of granulite-facies rocks from the Fraser Range, Western Australia, and the Musgrave Ranges, Central Australia: Geol. Soc. Australia Spec. Pub. 2, p. 377–388.

Assunçao, C. F. Torre de, and Coelho, A. V. P., 1956, The rocks with charnockitic affinities in Mozambique: Moçambique Serviços Indústria e Geologia Bol., v. 19, p. 173–181.

Aubague, M., and Hausknecht, J.-J., 1957, Nouvelle interprétation du socle sur la feuille Libreville (est et ouest): French Equatorial Africa, Dir. Mines et Géologie Bull., v. 8, p. 52–59.

——1959, Carte géologique de reconnaissance de l'Afrique Equatoriale Française au 1/500,000. Feuille Libreville-est, avec notice explicative: French Equatorial Africa, Dir. Mines et Géologie, 34 p.

Bagnall, P. S., 1963, The geology of the North Pare Mountains: Tanganyika Geol. Survey Recs., v. 10, p. 7–16.

——1964, Geological relationships in N.E. Tanganyika, and their bearing on the granulite problem: Univ. Leeds, Research Inst. African Geology, 8th Ann. Rept., p. 35–36.

Bagnall, P. S., Dundas, P. L., and Hartley, E. W., 1965, The geology of the Mnazi and Lushoto areas; quarter-degree sheets 90 and 109 and parts of 73E: Tanganyika Geol. Survey Recs., v. 11, p. 15–17.

Bahneman, K. P., 1971, *in* Morrison, E. R., and Wilson, J. F., Symposium on the granites, gneisses, and related rocks: Excursion guidebook: Geol. Soc. South Africa, Rhodesia Branch, p. 16–22.

Bahneman, K. P., 1972, A review of the structure, the stratigraphy and the metamorphism of the basement-rocks in the Messina district, northern Transvaal [D.Sc. thesis]: Pretoria, South Africa, Univ. Pretoria, 156 p.

Baker, B. H., 1963, Geology of the Baragoi area: Kenya Geol. Survey Rept. 53, 74 p.

Barrère, J., 1967, Le groupe Précambrien de l' Amsaga entre Atar et Akjoujt (Mauritanie). Étude d'un métamorphisme profond et de ses relations avec la migmatisation: [France] Bur. Recherches Géol. et Minières Mem. 42, 275 p.

——1969, Aperçu sur le métamorphisme et la migmatisation dans le Précambrien de l'Amsaga (Mauritanie sud-occidentale): Soc. Géol. France Bull., v. 11, sér. 7, p. 150–159.

Bates, D. A., 1957, The geological evolution of the Gold Coast: Comm. Tech. Co-operation Africa south of the Sahara, West-Central Regional Comm. Geology, 2d, Accra 1957, Proc., p. 13–22.

Bear, L. M., 1952, A geological reconnaissance of the area southeast of Embu: Kenya Geol. Survey Rept. 23, 39 p.

——1955, Geology of the Taveta area: Kenya Geol. Survey Rept. 32, 48 p.

Behr, H. J., and others, 1971, Granulites — Results of a discussion: Neues Jahrb. Mineralogie Monatsh., v. 3, p. 97–123.

Benedict, P. C., Wiid, D. de N., Cornelissen, A. K., and others, 1964, Progress report on the geology of the O'okiep Copper District, *in* Haughton, S. H., ed., The geology of some ore deposits of southern Africa, Vol. II: Johannesburg, Geol. Soc. South Africa, p. 239–302.

Bertrand, J.M.L., 1967, Existence de plissement superposés dans le Précambrien de l'Aleksod (Ahaggar oriental): Soc. Géol. France Bull., v. 9, sér. 7, p. 741–749.

Bessoles, B., 1955, Notice explicative sur la feuille Yalinga-ouest; Afrique équatoriale française, carte géologique de reconnaissance à l'échelle du 1/500,000: Paris, French Equatorial Africa, Dir. Mines et Géol., 24 p.

——1962, Géologie de la région de Bria et d'Ippy (République Centrafricaine). Contribution à l'étude de la migmatisation: [France] Bur. Recherches Géol. et Minières Mem. 18, 205 p.

Bessoles, B., and Roques, M., 1959, Âges apparents par la méthode plomb-alpha de zircons extraits de roches cristallines d'Afrique equatoriale française et du Cameroun: Internat. Geol. Cong., 20th, Mexico D. F. 1956, Asoc. Servicios Geol. Africanos, Actas y Trabajos, p. 35–37.

Beugnies, A., 1953, Le complexe de roches magmatiques de l'entre Lubilash-Lubishi (Katanga): Inst. Royal Colonial Belge, Bull., Sci. Tech., Mém., coll. in-8°, v. 23, no. 1, 128 p.

Black, R., 1967, Sur l'ordonnance des chaînes métamorphiques en Afrique occidentale: Chronique Mines et Recherche Minière, v. 364, p. 225–238.

Black, R., and Girod, M., 1970, Late Palaeozoic to Recent igneous activity in West Africa and its relationship to basement structure, *in* Clifford, T. N., and Gass, I. G., eds., African magmatism and tectonics: Edinburgh, Oliver and Boyd, p. 185–210.

Blanchot, A., 1954, Aperçu sur le Précambrien de Mauritanie occidentale: Internat. Geol. Cong., 19th, Algiers 1952, Assoc. Services Géol. Africains, pt. 1, p. 97–113.

——1955, Le Précambrien de Mauritanie occidentale: French West Africa, Dir. Mines et Géol. Bull. 17, 215 p.

Bloomfield, K., 1968, The pre-Karroo geology of Malawi: Malawi Geol. Survey. Mem. 5, 166 p.

Boissonnas, J., Borsi, S., Ferrara, G., Fabre, J., Fabries, J., and Gravelle, M., 1969, On the early Cambrian age of two late orogenic granites from west-central Ahaggar (Algerian Sahara): Canadian Jour. Earth Sci., v. 6, p. 25–37.

Bolgarsky, M., 1950, Étude Géologique et description pétrographique du sud-ouest de la Côte d'Ivoire: French West Africa, Dir. Mines Bull., v. 9, 170 p.

Bonhomme, M., 1962, Contribution à l'étude géochronologique de la plate-forme de l'Ouest-Africain: Clermont Univ. Fac. Sci. Annales, Géologie et Minéralogie, v. 5, no. 5, 62 p.

Borges, A., and Coelho, A.V.P., 1957, Primeiro reconhecimento petrográficoda circumscrição do Barué: Moçambique Serviços Indústria e Geologia Bol., v. 21, 80 p.

Caby, R., 1970, La chaîne pharusienne dans le nord-ouest de l'Ahaggar (Sahara central, Algérie); sa place dans l'orogenèse du Précambrien supérieur en Afrique [D.Sc. thesis]: Montpellier, France Univ. Montpellier 336 p.
Cahen, L., 1970, Igneous activity and mineralisation episodes in the evolution of the Kibaride and Katangide orogenic belts of Central Africa, *in* Clifford, T. N., and Gass, I. G., eds., African magmatism and tectonics: Edinburgh, Oliver and Boyd, p. 97–117.
Cahen, L., and Lepersonne, J., 1967, The Precambrian of the Congo, Rwanda, and Burundi, *in* Rankama, K., ed., The geologic systems, Vol. 3: New York, Interscience Pubs., p. 143–290.
Cahen, L., and Snelling, N. J., 1966, The geochronology of equatorial Africa: Amsterdam, North-Holland Pub. Co., 195 p.
Cannon, R. T., Hopkins, D. A., Thatcher, E. C., Peters, E. R., Kemp, J., Gaskell, J. L., and Ray, G. E., 1969, Polyphase deformation in the Mozambique Belt, northern Malawi: Geol. Soc. America Bull., v. 80, p. 2615–2622.
Carter, J. D., Barber, W., and Tait, E. A., 1963, The geology of parts of Adamawa, Bauchi and Bornu Provinces in northeastern Nigeria: Nigeria Geol. Survey Bull., v. 30, 109 p.
Champetier de Ribes, G., and Aubague, M., 1956, Notice Explicative sur la feuille Yaoundé-ouest; Territoire de Cameroun, carte géologique de reconnaissance à l'échelle du 1/500,000: Cameroun Dir. Mines et Géol., 35 p.
Champetier de Ribes, G., and Reyre, D., 1959, Notice explicative sur la feuille Yaoundé-ouest, Territoire de Cameroun, carte géologique de reconnaissance à l'échelle du 1/500,000: Cameroun Dir. Mines et. Géol., 31 p.
Chinner, G. A., and Sweatman, T. R., 1968, A former association of enstatite and kyanite: Mineralog. Mag., v. 36, p. 1052–1060.
Choubert, B., 1954, Recherches Géologiques au Gabon central: French Equatorial Africa, Dir. Mines et Géol. Bull., v. 6, p. 5–81.
Clifford, T. N., 1967, The Damaran episode in the upper Proterozoic–lower Paleozoic structural history of southern Africa: Geol. Soc. America Spec. Paper 92, 100 p.
——1968, Radiometric dating and the pre-Silurian geology of Africa, *in* Hamilton, E. I., and Farquhar, R. M., eds., Radiometric dating for geologists: New York, Interscience Pubs., p. 299–416.
——1970, The structural framework of Africa, *in* Clifford, T. N., and Gass, I. G., eds., African magmatism and tectonics: Edinburgh, Oliver and Boyd, p. 1–26.
——1972, The evolution of the African crust: Maroc Service Géol. Notes et Mém., v. 236, p. 29–39.
——1973, African granulites and related rocks: A preliminary note: Geol. Soc. South Africa Spec. Pub. 3, p. 17–24.
Clifford, T. N., Gronow, J., Rex, D. C. and Burger, A. J., 1974, Geochronological and petrogenetic studies of high-grade metamorphic rocks and intrusives in Namaqualand, South Africa: Jour. Petrology (in press).
Coelho, A.V.P., 1954, Quelques aspects de le pétrographie de la circonscription du Barué (Mozambique): Internat. Geol. Cong., 19th, Algiers 1952, Assoc. Services Géol. Africains, pt. 1, p. 275–292.
——1957, Anorthositos de Tete (Moçambique): Moçambique Serviços Indústria e Geologia, v. 24, p. 69–80.
——1969, O complexo gabro-anorthosítico de Tete (Moçambique): Moçambique Serviços Geologia e Minas Bol., v. 35, p. 63–78.
Coetzee, C. B., 1941, The petrology of the Goodhouse-Pella area in Namaqualand, South Africa: Geol. Soc. South Africa Trans., v. 44, p. 167–206.
——1942, Metamorphosed sediments from the Goodhouse-Pella area, Namaqualand, South Africa: Royal Soc. South Africa Trans., v. 29, p. 91–112.
Cox, K. G., Johnson, R. L., Monkman, L. J., Stillman, C. J., Vail, J. R., and Wood, D. N., 1965, The geology of the Nuanetsi igneous province: Royal Soc. London Philos. Trans., v. 257, ser. A, p. 71–218.

Crawford, A. R., 1969, Reconnaissance Rb-Sr dating of the Precambrian rocks of southern peninsular India: Geol. Soc. India Jour., v. 10, p. 117–166.

Daly, R. A., 1933, Igneous rocks and the depths of the earth: New York, McGraw-Hill Book Co., 508 p.

Darnley, A. G., Horne, J.E.T., Smith, G. H., Chandler, T.R.D., Dance, D. F., and Preece, E. R., 1961, Ages of some uranium and thorium minerals from east and central Africa: Mineralog. Mag. v. 32, p. 716–724.

Davidson, C. F., and Bennett, J.A.E., 1950, The uranium deposits of the Tete District, Mozambique: Mineralog. Mag., v. 29, p. 291–303.

Davies, R. D., Allsopp, H. L., Erlank, A. J., and Manton, W. I., 1970, Sr-isotopic studies on various layered mafic intrusions in southern Africa: Geol. Soc. South Africa Spec. Pub. 1, p. 576–593.

Delafosse, R., 1960, Notice explicative sur la feuille Ouanda-Djalle-ouest, République afrique centrale, carte géologique de reconnaissance à l'échelle du 1/500,000: French Equatorial Africa, Dir. Mines et Géologie, Inst. Equatoriale Recherches Etudes Géol. et Min., 38 p.

Delhal, J., 1957, Les massifs cristallins de la Lulua et de la Lueta (Kasai): Univ. Louvain Inst. Géol. Mém. 20, p. 211–281.

——1958, Les roches charnockitiques du Kasai (Congo Belge): Comm. Tech. Co-operation Africa south of Sahara, East-Central, West-Central, and South Regional Comm. Geology, joint mtg., Leopoldville 1958, Proc., p. 271–281.

——1963, Le socle de la région de Luiza (Kasai): Annales Mus. Royal de l'Afrique Centrale, v. 45, 82 p.

Delhal, J., and Fieremans, C., 1964, Extension d'un grand complexe charnockitique en Afrique centrale: Acad. Sci. Comptes Rendus, v. 259, p. 2665–2668.

Delhal, J., and Ledent, D., 1965, Quelques résultats géochronologiques relatifs aux formations du socle de la région de Luiza (Kasai): Soc. Géol. Belgique Bull., v. 74, p. 102–113.

——1971, Ages U/Pb et Rb/Sr et rapports initiaux du strontium du complexe gabbro-noritique et charnockitique du bouclier du Kasai (République Démocratique du Congo et Angola): Soc. Géol. Belgique Annales, v. 94, p. 211–221.

Dévigne, J. P., 1958, Le Précambrien du Gabon occidental en Afrique equatoriale française et les régions limitrophes: French Equatorial Africa, Dir. Mines et Géologie Bull., v. 11, 315 p.

De Villiers, J., and Söhnge, P. G., 1959, The geology of the Richtersveld: South Africa Geol. Survey Mem. 48, 295 p.

Dodson, R. G., 1963, Geology of the south Horr area: Geol. Survey Kenya Rept. 60, 53 p.

Drysdall, A.R., Johnson, R. L., Moore, T. A., and Thieme, J. G., 1972, Outline of the geology of Zambia: Geologie en Mijnbouw, v. 51, p. 265–276.

Dumort, J. C., 1968, Notice explicative sur la feuille Doualo-ouest, Territoire du Cameroun, carte géologique de reconnaissance à l'échelle du 1/500,000: French Equatorial Africa, Dir. Mines et Géologie, Inst. Equatoriale Recherches Etudes Géol. et Min., 69 p.

Eberhardt, P., Ferrara, G., Glangeaud, L., Gravelle, M., and Tongiorgi, E., 1963, Sur l'âge absolu des séries métamorphiques de l'Ahaggar occidental dans la région de Silet-Tibehaouine (Sahara central): Acad. Sci. Comptes Rendus, v. 256, p. 1126–1128.

Evans, C. R., 1965, Geochronology of the Lewisian basement near Lochinver, Sutherland: Nature, v. 207, p. 54–56.

Evans, R. K., 1965, The geology of the Shire Highlands: Malawi Geol. Survey Bull., v. 18, 54 p.

Faure, G., and Hurley, P. M., 1963, The isotopic composition of strontium in oceanic and continental basalts; application to the origin of igneous rocks: Jour. Petrology, v. 4, p. 30–50.

Ferragne, A., 1964, Caractères pétrographiques de la série charnockitique de Gour Oumelalen: Travaux Inst. Recherches Saharienne, v. 23, p. 139–152.

Ferrara, G., and Gravelle, M., 1966, Radiometric ages from western Ahaggar (Sahara) suggesting an eastern limit for the West African craton: Earth and Planetary Sci. Letters, v. 1, p. 319–324.

Fitches, W. R., 1968, New K/Ar age determinations from the Precambrian Mafingi Hills area of Zambia and Malawi: Univ. Leeds, Research Inst. African Geology, 12th Ann. Rept., p. 12–14.
——1970, A part of the Ubendian orogenic belt in northern Malawi and Zambia: Geol. Rundschau, v. 59, p. 444–458.
Fozzard, P.M.H., 1958, Geology of the Nachingwea area in the Southern Province of Tanganyika: Tanganyika Geol. Survey Recs., v. 6, p. 8–13.
Freitas, A. J. de, 1957, Noticia explicativa do esboço geologico de Moçambique (1:2,000,000): Moçambique Serviços Indústria e Geologia Bol., v. 23, 82 p.
Füster, J. M., 1958, Sur les variations de composition chimique pendant le métamorphisme dans le faciès des granulites: Geochim. et Cosmochim. Acta, v. 14, p. 154–155.
Gazel, J., Hourcq, V., and Nicklès, M., 1956, Carte géologique du Cameroun au 1/1,000,000, avec notice explicative: Cameroun Dir. Mines et Géologie Bull. 2, 62 p.
Gérard, G., and Gérard, J., 1953, Notice explicative sur la feuille Berbérati-est. Afrique equatoriale française, carte géologique de reconnaissance à l'échelle du 1/500,000: Paris, French Equatorial Africa, Dir. Mines et Géologie, 27 p.
Gérard, J., 1963, Notice explicative sur la feuille Bossangoa-est. Republique Centrafricaine, carte géologique de reconnaissance à l'échelle du 1/500,000: Central African Republic, Dir. Mines et Géologie, 61 p.
Gevers, T. W., and Dunne, J. C., 1942, Charnockitic rocks near Port Edward in Alfred County, Natal: Geol. Soc. South Africa Trans., v. 45, p. 183–213.
Gevers, T. W., Partridge, F. C., and Joubert, G. K., 1937, The pegmatite area south of the Orange River in Namaqualand: South Africa Geol. Survey Mem., v. 31, 180 p.
Giraud, P., 1961, Les charnockites et les roches associées du Suggarien à faciès In Ouzzal (Sahara Algérien): Soc. Géol. France Bull., v. 3, sér. 7, p. 165–170.
——1964, Les roches à caractère charnockitique de la série d'In Ouzzal en Ahaggar (Sahara central). Essai de nomenclature des séries charnockitiques: Internat. Geol. Cong., 22d, New Delhi 1964, Comptes Rendus, pt. 13, p. 1–20.
Grant, N. K., 1970, Geochronology of Precambrian basement rocks from Ibadan, southwestern Nigeria: Earth and Planetary Sci. Letters, v. 10, p. 29–38.
Gray, C. M., 1971, Strontium isotopic studies on granulites [Ph.D. thesis]: Canberra, Australian National Univ., 248 p.
Green, D. H., and Ringwood, A. E., 1967, An experimental investigation of the gabbro to eclogite transformation and its petrological implications: Geochim. et Cosmochim. Acta, v. 31, p. 767–833.
Green, T. H., 1969, High-pressure experimental studies on the origin of anorthosite: Canadian Jour. Earth Sci., v. 6, p. 427–440.
Groves, A. W., 1935, The charnockitic series of Uganda, British East Africa: Geol. Soc. London Quart. Jour., v. 91, p. 150–207.
Harpum, J. R., 1954, Some problems of the pre-Karroo geology in Tanganyika: Internat. Geol. Cong., 19th, Algiers 1952, Assoc. Services Géol. Africains, pt. 20, p. 209–239.
Haughton, S. H., 1963, The stratigraphic history of Africa south of the Sahara: Edinburgh, Oliver and Boyd, 365 p.
Heier, K. S., 1964, Rubidium-strontium and strontium-87/strontium-86 ratios in deep crustal material: Nature, v. 202, p. 477–478.
Heier, K. S., and Compston, W., 1969, Interpretation of Rb-Sr age patterns in high-grade metamorphic rocks, north Norway: Norsk. Geol. Tidsskr., v. 49, p. 257–283.
Heier, K. S., and Thoresen, K., 1971, Geochemistry of high grade metamorphic rocks, Lofoten-Vesterålen, north Norway: Geochim. et Cosmochim. Acta, v. 35, p. 89–99.
Hensen, B. J., and Green, D. H., 1970, Experimental data on coexisting cordierite and garnet under high grade metamorphic conditions: Physics Earth and Planetary Interiors, v. 3, p. 431–440.

Hepworth, J. V., 1964a, Explanation of the geology of sheets 19, 20, 28 and 29 (southern West Nile): Uganda Geol. Survey Rept. 10, 123 p.
Hepworth, J. V., 1964b, The charnockites of southern West Nile, Uganda, and their parageneses: Internat. Geol. Cong. 22d, New Delhi 1964, Comptes Rendus, pt. 13, p. 168–184.
——1967, The photogeological recognition of ancient orogenic belts in Africa: Geol. Soc. London Quart. Jour., v. 123, p. 253–292.
Hepworth, J. V., and Macdonald, R., 1966, Orogenic belts of the northern Uganda basement: Nature, v. 210, p. 726–727.
Hirschberg, A., and Winkler, H.G.F., 1968, Stabilitäts-beziehungen zwischen Chlorit, Cordierit und Almandin bei der Metamorphose: Contr. Mineralogy and Petrology, v. 18, p. 17–42.
Holdaway, M. J., 1971, Stability of andalusite and the aluminum silicate phase diagram: Am. Jour. Sci., v. 271, p. 97–131.
Holland, T. H., 1900, The charnockite series, a group of Archaean hypersthenic rocks in Peninsular India: India Geol. Survey Mem. 28, pt. 2, 158 p.
Holt, D. N., 1961, The geology of part of the Fort Johnston District east of Lake Nyasa: Nyasaland Geol. Survey Rec., v. 1, p. 23–39.
Howie, R. A., 1958, African charnockites and related rocks: Belgian Congo, Service Géol. Bull., v. 8, no. 2, p. 1–14.
Hudeley, H., and Belmonte, Y., 1970, Carte géologique de la République Gabonaise au 1/1,000,000, notice explicative: [France] Bur. Recherches Géol. et Minières Mém. 72, 192 p.
Hunter, D. R., 1970, The Ancient Gneiss Complex in Swaziland: Geol. Soc. South Africa Trans., v. 73, p. 107–150.
Hurley, P. M., Rand, J. R., Fairbairn, H. W., Pinson, W. H., Posadaz, V. G., and Reid, J. B., 1966, Continental drift investigations: Massachusetts Inst. Technology, Dept. Geology and Geophysics, 14th Ann. Prog. Rept., p. 3–15.
Hurley, P. M., Leon, G. W., White, R. W., and Fairbairn, H. W., 1971, Liberian age province (about 2,700 m.y.) and adjacent provinces in Liberia and Sierra Leone: Geol. Soc. America Bull., v. 82, p. 3483–3490.
Jäger, E., and Watznauer, A., 1969, Einige Rb/Sr-Datierungen an Granuliten des sächsischen Granulitgebirges: Deutsch. Akad. Wiss. Berlin Monatsber., v. 11, p. 420–426.
James, T. C., 1951, Degree sheets 85, 86, 92 and 93—Songea District: Tanganyika Geol. Survey Ann. Rept. for 1949, p. 30–31.
——1958, Geological mapping in the Songea district: Progress report: Tanganyika Geol. Survey Rept. 6, p. 32.
Jansen, H., 1960, The geology of the Bitterfontein area, Cape province. An explanation of sheet 253 (Bitterfontein): Pretoria, South Africa Geol. Survey, 84 p.
Jones, H. A., and Hockey, R. D., 1964, The geology of part of southwestern Nigeria: Nigeria Geol. Survey Bull., v. 31, 101 p.
Joubert, P., 1966, Geology of the Loperot area: Kenya Geol. Survey Rept. 74, 52 p.
——1971, The regional tectonism of the gneisses of part of Namaqualand: Cape Town Univ. Dept. Geology Chamber Mines Precambrian Research Unit Bull. 10, 220 p.
Junner, N. R., 1940, Geology of the Gold Coast and western Togoland, with revised geological map: Gold Coast Geol. Survey Bull., v. 11, 40 p.
Junner, N. R., and Bates, D. A., 1945, Reports on the geology and hydrology of the coastal area east of the Akwapim Range: Gold Coast Geol. Survey Mem. 7, 23 p.
Karpoff, R., 1958, La géologie de l'Adrar des Iforas (Sahara central) [D.Sc. thesis]: Paris, Univ. Paris Fac. Sci., 271 p.
Kazmin, V., 1972, Granulites in Ethiopian basement: Nature, v. 240, p. 90–92.
Kennedy, W. Q., 1965, The influence of basement structure on the evolution of the coastal (Mesozoic and Tertiary) basins of Africa, *in* Salt basins around Africa: London, Institute of Petroleum, p. 7–16.

Kröner, A., 1971, The origin of the southern Namaqualand gneiss complex, South Africa, in the light of geochemical data: Lithos, v. 4, p. 325–344.
Kuno, H., 1968, Differentiation of basalt magmas, *in* Hess, H. H., and Poldervaart, A., eds., Basalts—The Poldervaart treatise on rocks of basaltic composition, Vol. 2: New York, Interscience Pubs., p. 623–688.
Lacroix, A., 1910, Sur l'existence à la Côte d'Ivoire d'une série pétrographique comparable à celle de la charnockite: Acad. Sci. Comptes Rendus, v. 150, p. 18.
Lambert, I. B., and Heier, K. S., 1968, Geochemical investigations of deep-seated rocks in the Australian Shield: Lithos, v. 1, p. 30–53.
Lasserre, M., 1964a, Étude géochronologique par la méthode strontium-rubidium de quelques echantillons en provenence du Cameroun: Clermont Univ. Fac. Sci. Annales, Géologie et Minéralogie, v. 25, p. 53–67.
——1964b, Mesures d'âges absolus sur les séries précambriennes et paléozoiques du Cameroun (Afrique équatoriale): Acad. Sci. Comptes Rendus, v. 258, p. 998–1000.
Lay, C., and Ledent, D., 1963, Mesures d'âges absolus de minéraux et de roches du Hoggar (Sahara central): Acad. Sci. Comptes Rendus, v. 257, p. 3188–3191.
Lay, C., Ledent, D., and Grögler, N., 1965, Mesures d'âges absolus de zircons du Hoggar (Sahara central) par la méthode uranium/plomb: Acad. Sci. Comptes Rendus, v. 260, p. 3113–3115.
Le Fur, Y., 1966, Nouvelles observations sur la structure de l'antécambrien du Hoggar nord occidental (Région d'In Hihaou) [D.Sp. thesis]: Nancy, France, Univ. Nancy Fac. Sci., 109 p.
Leggo, P. J., Aftalion, M., and Pidgeon, R. T., 1971, Discordant zircon U-Pb ages from the Uganda basement: Nature, v. 231, p. 81–84.
Legoux, P., 1939, Le Massif de Man (Côte d'Ivoire). Essai de géologie pétrographique: French West Africa, Service Mines Bull. 3, 92 p.
Lelubre, M., 1952a, L'antécambrien de l'Ahaggar (Sahara central): Internat. Geol. Cong., 19th, Algiers 1952, Mon. Régionale, sér. 1, no. 6, 147 p.
——1952b, Recherches sur la géologie de l'Ahaggar central et occidental (Sahara central): Algeria, Service Carte Géol. Bull. 22, sér. 2, 354 p.
——1969, Chronologie du Précambrien au Sahara central: Geol. Assoc. Canada Spec. Paper 5, p. 27–32.
Liebenberg, C. J., 1969, The trace elements of the Bushveld Igneous Complex, *in* Danchin, R. V., and Ferguson, J., The geochemistry of the Losberg intrusion, Fochville, Transvaal: Geol. Soc. South Africa Spec. Pub. 1, p. 689–714.
Macdonald, G. A., 1968, Composition and origin of Hawaiian lavas, *in* Coats, R. R., Hay, R. L., and Anderson, C. A., eds., Studies in volcanology (Williams volume): Geol. Soc. America Mem. 116, p. 477–522.
Macdonald, G. A., and Katsura, T., 1964, Chemical composition of Hawaiian lavas: Jour. Petrology, v. 5, p. 82–133.
Macdonald, R., 1964, "Charnockites" in the West Nile District of Uganda: A systematic study of Groves' type area: Internat. Geol. Cong., 22d, New Delhi 1964, Comptes Rendus, pt. 13, p. 227–249.
——1966, Explanatory note, and geological map of Uganda, *in* The atlas of Uganda: Entebbe, Uganda Dept. Lands Survey, scale 1:1,500,000.
Macgregor, A. M., 1951, Some milestones in the Precambrian of Southern Rhodesia: Geol. Soc. South Africa Trans. and Proc., v. 54, p. xxvii–lxxi.
——1952, Precambrian formations of tropical southern Africa: Internat. Geol. Cong., 19th, Algiers 1952, pt. 1, p. 39–50.
MacGregor, J. P., 1962, Explanation of the geology of sheet 10 (Kaabong): Uganda Geol. Survey Rept. 9, 42 p.
Marvier, L., 1953, Notice explicative de la carte géologique d'ensemble de l'Afrique occidentale française: French West Africa, Dir. Mines Bull. 16, 104 p.

Mason, R., 1969, Transcurrent dislocation in the Limpopo orogenic belt: Geol. Soc. London Proc., no. 1655, p. 93–96.
Matheson, F. J., 1966, Geology of the Kajiado area: Kenya Geol. Survey Rept. 70, 51 p.
McConnell, R. B., 1950, Outline of the geology of Ufipa and Ubende: Tanganyika Geol. Survey Bull., v. 19, 62 p.
McIver, J. R., 1966, Orthopyroxene-bearing granitic rocks from southern Natal: Geol. Soc. South Africa Trans. and Proc., v. 69, p. 99–117.
Mestraud, J. L., 1964, Carte géologique de la République Centrafricaine au 1:1,500,000: [France] Bur. Recherches Géol. et Minières, 1 sheet.
——1974, Géologie et ressources minérales de la République Centrafricaine: [France] Bur. Recherches Géol. et Minières Mem. 60 (in press).
Miller, J. M., 1956, Geology of the Kitale-Cherangani Hills area: Kenya Geol. Survey Rept. 35, 34 p.
Nicolaysen, L. O., 1962, Stratigraphic interpretation of age measurements in southern Africa, *in* Engel, A.E.J., James, H. L., and Leonard, B. F., eds., Petrologic studies (Buddington volume): New York, Geol. Soc. America, p. 569–598.
Nicolaysen, L. O., and Burger, A. J., 1965, Note on an extensive zone of 1000 million-year old metamorphic and igneous rocks in southern Africa: Sci. Terre, v. 10, nos. 3–4, p. 500–518.
Nixon, P. H., Reedman, A. J., and Burns, L. K., 1973, Sapphirine-bearing granulites from Labwor, Uganda: Mineralog. Mag., v. 39, p. 420–428.
Noizet, G., 1969, Contribution à l'étude géochimique des formations métamorphiques du faciès granulite dans le sud de Madagascar [D.Sc. thesis]: Nancy, France, Univ. Nancy, 188 p.
Oberholzer, W. F., compiler, 1968, Carta geológica do Moçambique: Direcção Provincial Serviços Geologia e Minas, scale 1:2,000,000, 1 sheet.
Oliver, R. L., 1969, Some observations on the distribution and nature of granulite-facies terrains: Geol. Soc. Australia Spec. Pub. 2, p. 259–268.
Oyawoye, M. O., 1961, On an occurrence of fayalite quartz monzonite in the basement complex around Bauchi, northern Nigeria: Geol. Mag., v. 98, p. 473–482.
——1964, The geology of the Nigerian Basement Complex: Nigerian Mining, Geol. and Metall. Soc. Jour., v. 1, p. 87–102.
Papon, A., Roques, M., and Vachette, M., 1968, Age de 2700 millions d'années, déterminé par la méthode au strontium, pour la série charnockitique de Man, en Côte d' Ivoire: Acad. Sci. Comptes Rendus, v. 266, p. 2046–2048.
Picciotto, E., Ledent, D., and Lay, C., 1965, Étude géochronologique de quelques roches du socle cristallophyllien du Hoggar (Sahara central): Sci. Terre, v. 10, nos. 3–4, p. 481–495.
Pichamuthu, C. S., 1969, Nomenclature of charnockites: Indian Minerals, v. 10, p. 23–35.
Poldervaart, A., 1966, Archaean charnockitic adamellite phacoliths in the Keimoes-Kakamas region, Cape Province, South Africa: Geol. Soc. South Africa Trans. and Proc., v. 69, p. 139–154.
Poldervaart, A., and Von Backström, J. W., 1949, A study of an area at Kakamas (Cape Province): Geol. Soc. South Africa Trans. and Proc., v. 52, p. 433–495.
Pollett, J. D., 1951, Geology and mineral resources of Sierra Leone: Colonial Geol. and Mineral Resources Bull., v. 2, p. 3–28.
Pougnet, R., 1957, Le Précambrien du Dahomey: French West Africa, Dir. Fédérale Mines et Géologie Bull., v. 22, 191 p.
Pouit, G., 1958, Sur la presence de séries charnockitiques à faciès granulite en Oubangui-central: Comm. Tech. Co-operation Africa south of the Sahara, East-Central, West-Central, and South Regional Comm. Geology, joint mtg., Leopoldville 1958, Proc., p. 285–290.
——1959, Étude géologique des formations métamorphiques, granitiques et charnockitiques de la région de Fort Crampel (Oubangui-Chari): French Equatorial Africa, Dir. Mines et Géol. Bull. 13, 144 p.
Pulfrey, W., 1946, A suite of hypersthene-bearing plutonic rocks in the Meru District, Kenya: Geol. Mag., v. 83, p. 67–88.

Quennell, A. M., McKinlay, A.C.M., and Aitken, W. G., 1956, Summary of the geology of Tanganyika: Tanganyika Geol. Survey Mem. 1, pt. 2, 264 p.

Radier, H., 1959, Contribution à l'étude géologique du Soudan oriental (A.O.F.); Vol. 1, Le Précambrien saharien au sud de l'Adrar des Iforas: French West Africa, Service Géol. Bull. 26, 305 p.

Reboul, C., Moussu, H., and Lessard, L., 1962, Notice explicative de la carte géologique au 1/500,000 du Hoggar (Sahara central): [France] Bur. Recherches Géol. et Minières, 96 p.

Ribeiro de Albuquerque, C. A., and Figueiredo Gomes, C. S., 1962, Rochas do corte da estrada Quizenga-Lucala-Sambacajú-Salazar (Angola)—Rochas de carácter charnoquítico: Coimbra Univ. Mus. e Lab. Mineralog. e Geol. Mem. e Notícias, v. 53, p. 53–73.

Richardson, S. W., 1968, Staurolite stability in a part of the system Fe-Al-Si-O-H: Jour. Petrology, v. 9, p. 467–488.

Richardson, S. W., Gilbert, M. C., and Bell, P. M., 1969, Experimental determination of kyanite-andalusite and andalusite-sillimanite equilibria; the aluminum silicate triple point: Am. Jour. Sci., v. 267, p. 259-272.

Robertson, I.D.M., 1968, Granulite metamorphism of the basement complex in the Limpopo metamorphic zone: Geol. Soc. South Africa Trans., v. 71, annexure, p. 125–133.

Rocci, G., 1965, Essai d'interprétation de mesures géochronologiques. La structure de l'Ouest Africain: Sci. Terre, v. 10, nos. 3–4, p. 461–479.

Rogers, A. W., and Du Toit, A. L., 1908, Report on the geology of parts of Prieska, Hay, Britstown, Carnarvon and Victoria West: Geol. Comm. Cape of Good Hope Ann. Rept., v. 13, p. 8–127.

Roques, M., 1948, Le Précambrien de l'Afrique occidentale française: Soc. Géol. France Bull., v. 18, sér. 5, p. 589–628.

Roubault, M., Delafosse, R., Leutwein, F., and Sonet, J., 1965, Premières données géochronologiques sur les formations granitiques et cristallophylliennes de la République Centre-Africaine: Acad. Sci. Comptes Rendus, v. 260, p. 4787–4792.

Saggerson, E. P., 1957, Geology of the South Kitui area: Kenya Geol. Survey Rept. 37, 49 p.

Saggerson, E. P., and Owen, L. M., 1969, Metamorphism as a guide to depth of the top of the mantle in southern Africa: Geol. Soc. South Africa Spec. Pub. 2, p. 335–349.

Sampson, D. N., and Wright, A. E., 1964, The geology of the Uluguru Mountains: Tanzania Geol. Survey Bull., v. 37, 69 p.

Sanders, L. D., 1954, Geology of the Kitui area: Kenya Geol. Survey Rept. 30, 53 p.

Schoeman, J. J., 1951, A geological reconnaissance of the country between Embu and Meru: Kenya Geol. Survey Rept. 17, 57 p.

——1952, A suite of charnockitic rocks from the Meru District, Kenya Colony [D.Sc. thesis]: Johannesburg, Univ. Witwatersrand, 85 p.

Schreyer, W., and Yoder, H. S., 1964, The system Mg-cordierite-H_2O and related rocks: Neues Jahrb. Mineralogie Abh., v. 101, p. 271–342.

Searle, D. L., 1952, The geology of the area northwest of Kitale township (Trans Nzioa, Elgon and West Suk): Kenya Geol. Survey Rept., v. 19, 80 p.

Seifert, F., 1970, Low-temperature compatibility relations of cordierite in haplopelites of the system K_2O-MgO-Al_2O_3-SiO_2-H_2O: Jour. Petrology, v. 11, p. 73–99.

Sighinolfi, G. P., 1971, Investigations into deep crustal levels: Fractionating effects and geochemical trends related to high-grade metamorphism: Geochim. et Cosmochim. Acta, v. 35, p. 1005–1021.

Simpson, E.S.W., and Tregidga, J. A., 1956, The Archaean rocks of the Marble Delta District, Natal: Geol. Soc. South Africa Trans., v. 59, p. 237–258.

Snelling, N. J., 1966, Age determination unit: Inst. Geol. Sci. Ann. Rept. for 1965, Pt. II, Overseas Geol. Survey, p. 44–57.

Söhnge, P. G., Le Roex, H. D., and Nel, H. J., 1948, The geology of the country around Messina; an explanation of sheet no. 46 (Messina): South Africa Geol. Survey, 82 p.

Sougy, J., 1969, Grand lignes structurales de la chaîne des Mauritanides et de son avant-pays (socle précambrien et sa couverture infracambrienne et paléozoique), Afrique de l'Ouest: Soc. Géol. France Bull., v. 11, sér. 7, p. 133–149.

Spooner, C. M., 1969, Sr^{87}/Sr^{86} initial ratios and whole-rock ages of pyroxene granulites: Massachusetts Inst. Technology, Dept. Earth and Planetary Sciences, 17th Ann. Prog. Rept., p. 45–93.

Spooner, C. M., and Fairbairn, H. W., 1970, Strontium 87/strontium 86 initial ratios in pyroxene granulite terranes: Jour. Geophys. Research, v. 75, p. 6706–6713.

Spooner, C. M., Hepworth, J. V., and Fairbairn, H. W., 1970, Whole-rock Rb-Sr isotopic investigation of some East African granulites: Geol. Mag., v. 107, p. 511–521.

Steuber, A. M., and Murthy, V. R., 1966, Strontium isotope and alkali element abundances in ultramafic rocks: Geochim. et Cosmochim. Acta, v. 30, p. 1243–1259.

Stockley, G. M., 1948, Geology of north, west and central Njombe District, Southern Highland Province: Tanganyika Geol. Survey Bull., v. 18, 70 p.

Stowe, C. W., 1968, Intersecting fold trends in the Rhodesian Basement Complex south and west of Selukwe: Geol. Soc. South Africa Trans., v. 71, annexure, p. 53–78.

Sutton, J., Watson, J., and James, T. C., 1954, A study of the metamorphic rocks of Karema and Kungwe Bay, western Tanganyika: Tanganyika Geol. Survey Bull., v. 22, 70 p.

Teale, E. O., Eades, N. W., and Oates, F., 1935, The eastern Lupa Goldfield: Tanganyika Geol. Survey Bull., v. 8, 61 p.

Temperley, B. N., 1938, The geology of the country around Mpwapwa: Tanganyika Geol. Survey Short Paper 19, 61 p.

Trendall, A. F., 1961, Explanation of the geology of sheet 45 (Kadam): Uganda Geol. Survey Rept. 6, 46 p.

——1965a, Explanation of the geology of sheet 44 (Magoro): Uganda Geol. Survey Rept. 11, 28 p.

——1965b, Explanation of the geology of sheet 35 (Napak): Uganda Geol. Survey Rept. 12, 70 p.

Truswell, J. F., 1970, An introduction to the historical geology of South Africa: Cape Town, Purnell & Sons, 167 p.

Truswell, J. F., and Cope, R. N., 1963, The geology of parts of Niger and Zaria Provinces, northern Nigeria: Nigeria Geol. Survey Bull., v. 29, 53 p.

Tyrrell, G. W., 1916, A contribution to the petrography of Benguella, based on a rock collection made by Professor J. W. Gregory: Royal Soc. Edinburgh Trans., v. 51, p. 537–559.

Vachette, M., Razafiniparany, A., and Roques, M., 1969, Ages au strontium de 2700 millions d'années et de 1000 millions d'années, pour deux massifs charnockitiques de Madagascar: Acad. Sci. Comptes Rendus, v. 269, p. 1471–1473.

Vail, J. R., 1968, Preliminary investigation of the Tete Complex, Mozambique: Univ. Leeds, Research Inst. African Geology, 12th Ann. Rept., p. 23–25.

Van Breemen, O., 1970, Geochronology of the Limpopo orogenic belt, southern Africa: Jour. Earth Sci. (Leeds), v. 8, p. 57–61.

Van Niekerk, C. B., and Burger, A. J., 1969, A note on the minimum age of the acid lava of the Onverwacht Series of the Swaziland System: Geol. Soc. South Africa Trans., v. 72, p. 9–21.

Velde, B., 1966, Upper stability of muscovite: Am. Mineralogist, v. 51, p. 924–929.

Viljoen, M. J., and Viljoen, R. P., 1970, Archaean vulcanicity and continental evolution in the Barberton Region, Transvaal, *in* Clifford, T. N., and Gass, I. G., eds., African magmatism and tectonics: Edinburgh, Oliver and Boyd, p. 27–49.

Visser, D.J.L., compiler, 1956, The geology of the Barberton area: South Africa Geol. Survey Spec. Pub. 15, 253 p.

Von Backström, J. W., 1964, The geology of an area around Keimoes, Cape Province, with special reference to phacoliths of charnockitic adamellite-porphyry: South Africa Geol. Survey Mem. 53, 218 p.

——1967, The geology and mineral deposits of the Riemvasmaak area, northwest Cape Province: South Africa Geol. Survey Annals, v. 6, p. 43–51.

Von Knorring, O., and Kennedy, W. Q., 1958, The mineral paragenesis and metamorphic status of garnet-hornblende-pyroxene-scapolite gneiss from Ghana (Gold Coast): Mineralog. Mag., v. 31, p. 846–859.

Wacrenier, P., and Wolff, J. P., 1965, Notice explicative sur la feuille Bangui-ouest, Carte géologique de reconnaissance à l'échelle du 1/500,000: French Equatorial Africa, Dir. Mines et Géologie, Inst. Equatoriale Recherches Etudes Géol. et Minières, 56 p.

Wakefield, J., 1971, Preliminary report on the Pikwe area, Limpopo belt, Botswana: Univ. Leeds, Research Inst. African Geology, 15th Ann. Rept., p. 35–38.

Walsh, J., 1966, Geology of the Karasuk area: Kenya Geol. Survey Rept. 72, 34 p.

Walshaw, R. D., 1965, The geology of the Ncheu-Balaka area: Malawi Geol. Survey Bull., v. 19, 96 p.

Wickremasinghe, O. C., 1969, The geochemistry and geochronology of the charnockites and associated rocks of Ceylon [Ph.D. thesis]: Leeds, England, Univ. Leeds, 155 p.

Willemse, J., 1938, On the old granite of the Vredefort region and some of its associated rocks: Geol. Soc. South Africa Trans. and Proc., v. 40, p. 43–119.

Williams, L.A.J., 1966, Geology of the Chanler's Falls area: Kenya Geol. Survey Rept. 75, 54 p.

Wilson, J. F., and Harrison, N. M., 1973, Recent K-Ar age determinations on some Rhodesian granites: Geol. Soc. South Africa Spec. Pub. 3, p. 69–78.

Wilson, N. W., 1965, Geology and mineral resources of part of the Gola Forests, southeastern Sierra Leone: Sierra Leone Geol. Survey Bull., v. 4, 102 p.

Wolff, J. P., 1963, Notice explicative sur la feuille Yalinga-est. République Centrafricaine, Carte géologique de reconnaissance à l'échelle du 1/500,000: [France] Bur. Recherches Géol. et Minières, 38 p.

Wright, A. E., and James, T. C., 1958, Charnockites in Tanganyika and their associated rock groups: Comm. Tech. Co-operation Africa south of the Sahara, East-Central, West-Central, and South Regional Comm. Geology, joint mtg., Leopoldville 1958, Proc., p. 293–294.

MANUSCRIPT RECEIVED BY THE SOCIETY JUNE 4, 1973
REVISED MANUSCRIPT RECEIVED FEBRUARY 5, 1974